AMAZING MODELS!
WATER POWER

PETER HOLLAND

TAB BOOKS Inc.
Blue Ridge Summit, PA

FIRST EDITION
FIRST PRINTING

Copyright © 1990 in North America by TAB BOOKS Inc.
Printed and bound in Great Britain

Library of Congress Cataloging-in-Publication Data

Holland, Peter, 1951–
    Amazing models : water power/by Peter Holland.
        p.cm.
    ISBN 0-8306-3502-5 : $7.70
    1. Hydraulic models.   I. Title
TC164, h65   1990   89-29246
820.1'06--dc20   CIP

CONTENTS

Introduction

Model making is fun and it doesn't have to be expensive.

The aim of *Amazing Models!* is to show you how you can make working models – of, among others, a chicken, mystic fountain, aqua pod and clock – using only the most simple household materials. You will be amazed when you see what these models can do for little or no cost and with the minimum of assembly time.

In this volume, there are eight models and a chapter is devoted to each. Each chapter contains a drawing of the finished model, some simple instructions and a series of full-size, detailed drawings which you can copy onto the required materials.

The tools and materials you will need can probably be found around the house and, if not, can be bought cheaply at an art shop, DIY shop or at your local store.

WHAT YOU WILL NEED
A) Tools
— a small, very sharp modelling knife with a pointed blade
— scissors
— small 'snipe-nosed' pliers (choose the type which can cut as well)
— a metal straight edge – a piece of strip aluminium left over from some DIY job is ideal, or even a piece of hard plastic which has a really straight edge
— a ruler – if this has a hard edge that will not be cut by the modelling knife, then use it in place of the metal straight edge
— a school compass, which will hold a pencil. It may even hold the modelling knife, but that is not essential
— an HB pencil or fine tip felt pen

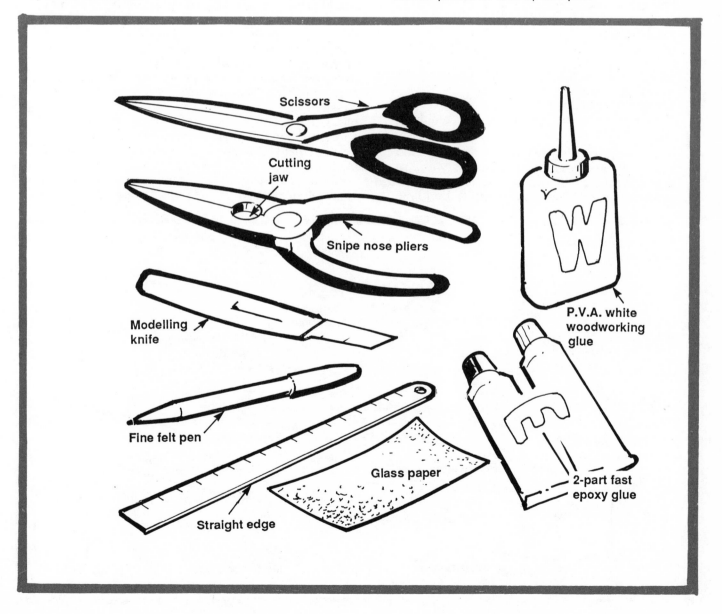

Scissors

Cutting jaw

Snipe nose pliers

P.V.A. white woodworking glue

Modelling knife

Fine felt pen

Straight edge

Glass paper

2-part fast epoxy glue

B) Materials
— expanded plain polystyrene ceiling tiles (about ¼″ [6mm] thick)
— balsa wood
— thin card (old postcards, cornflake packets etc.)
— corrugated cardboard (small pieces from old boxes)
— thin plastic cut from empty margarine containers – the containers themselves are also useful for some of the models
— empty yoghurt pots, especially those with a shaped foot which makes an ideal pulley wheel
— 35mm film cassette tubs
— empty ballpoint pen cases
— cotton buds with hollow stems (for tubes and bearings)
— plastic drinking straws (straight or bendy)
— 'giant' size wire paper clips for reshaping into wire parts
— wooden cocktail sticks
— pins
— modelling clay (plasticine)
— fine glasspaper
— sequins
— masking tape
— sewing thread, including 'invisible' mending nylon monofilament
— greaseproof paper (for copying shapes from the model drawings)
— a piece of hardboard or thick cardboard to protect the table when you are cutting and building

C) Glues
— white woodworking (PVA) glue can be used for the card, balsa and expanded polystyrene parts
— fast epoxy glue (a two part adhesive is needed when wire and plastic parts are to be joined to each other or to the other materials) – if you are in a hurry, use this throughout.
— faster still is super glue, but this melts the expanded polystyrene. It is not essential for these models.

When the models are built, you can decorate them with water paints. An alternative is to use poster paint mixed with emulsion paint. Remember that some other paints, particularly those containing oil or cellulose, melt the expanded polystyrene.

TIPS FOR SPEEDY MODELLING
Before starting on the models described in this book, it's a good idea to teach yourself wire bending, making bearings, joints and shapes, cut from various materials. This chapter shows you how, and it will speed you through the construction of each design.

Paper Clips
All the wire parts in this book are made by reshaping ordinary 'giant' size paper clips. These can be bought cheaply at office stationery shops and may also be found in good general stationery shops, but check the size with Fig. 1. They should be smooth rather than lipped or corrugated. Observe the length of the piece of wire when one is opened right out. Some parts can start at a straight part of the clip, to save reshaping. This is where the pliers come in. It is possible to shape wire parts with the fingers, but pliers make a more accurate job. Always grip the shorter end in the pliers, before bending the longer part to make the bend. The sequence in Fig. 2 explains this.

When you make a piece of wire into an axle, or some other part which has bends each end and a

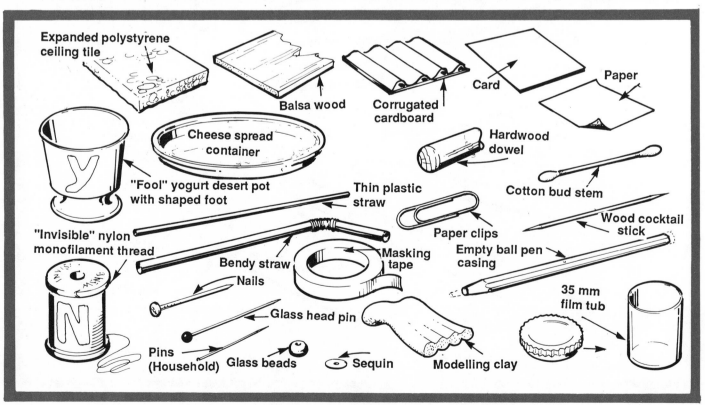

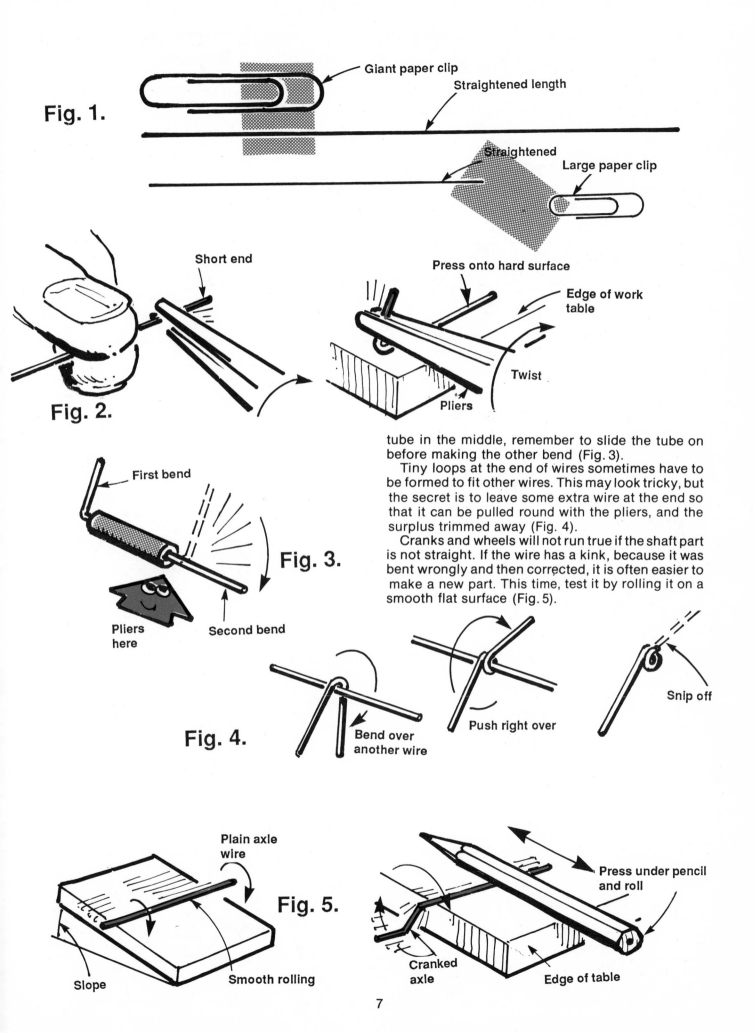

Fig. 1.

Giant paper clip

Straightened length

Straightened

Large paper clip

Fig. 2.

Short end

Press onto hard surface

Edge of work table

Twist

Pliers

Fig. 3.

First bend

Pliers here

Second bend

tube in the middle, remember to slide the tube on before making the other bend (Fig. 3).

Tiny loops at the end of wires sometimes have to be formed to fit other wires. This may look tricky, but the secret is to leave some extra wire at the end so that it can be pulled round with the pliers, and the surplus trimmed away (Fig. 4).

Cranks and wheels will not run true if the shaft part is not straight. If the wire has a kink, because it was bent wrongly and then corrected, it is often easier to make a new part. This time, test it by rolling it on a smooth flat surface (Fig. 5).

Fig. 4.

Bend over another wire

Push right over

Snip off

Fig. 5.

Plain axle wire

Press under pencil and roll

Slope

Smooth rolling

Cranked axle

Edge of table

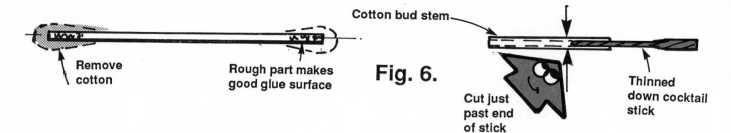

Remove cotton

Rough part makes good glue surface

Cotton bud stem

Fig. 6.

Cut just past end of stick

Thinned down cocktail stick

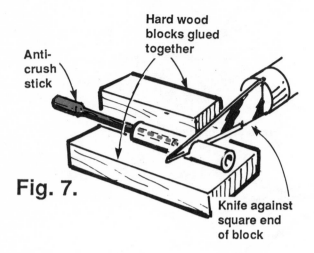

Hard wood blocks glued together

Anti-crush stick

Fig. 7.

Knife against square end of block

Bearings

To make these, cut cotton bud stems with a modelling knife but, to avoid squashing them, slip a piece of matchstick or the end of a wooden cocktail stick inside. Glasspaper or shave one down and keep it to slide inside the stems near to where the cut is to be made, as in Fig. 6. Make sure that the cut is at 90° to the length of the tube by holding the blade against a piece of hard wood (Fig. 7). After cutting, push the point of a pencil into the end and rotate it. This smoothes away any rough edges and leaves a coating of graphite to make the wire run smoothly. If any cotton remains on the stem, clean it away as it may tangle in the shaft and slow it.

When a bearing tube is to be glued into the model, there is a chance that the glue (epoxy) will get inside and clog the hole or, worse, stick to the shaft if it is slid in immediately afterwards. Look at Fig. 8 to see how cocktail stick ends seal the tube. A smear of modelling clay will also serve, but has to be cleaned out afterwards. Wax can also be used and is more easily cleaned away.

Expanded polystyrene tiles

For simplicity, these will be referred to in the instructions and on the drawings as just 'tiles'. If you have a miniature hot wire cutting tool it can be put to good use here, but most of the parts can be cut with the modelling knife held against a straight edge. Use several light slicing strokes. Do not use scissors. Fig. 9 shows how to cut circles for wheels.

It is important to protect the centre with masking tape or glue the card reinforcing piece in place first (most designs show such a card part). This ensures that the compass point does not skid, so forming the wrong radius, and gives the wheel a firm centre to carry a shaft or bearing tube. This also helps it to run true when the rim is finished on what is called a 'wheel lathe', seen in Fig. 10. This is a very useful item, for some models need really smooth edges to the wheels. Lumpy wheels absorb too much power.

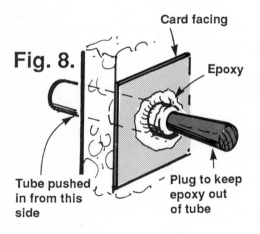

Fig. 8.

Card facing

Epoxy

Tube pushed in from this side

Plug to keep epoxy out of tube

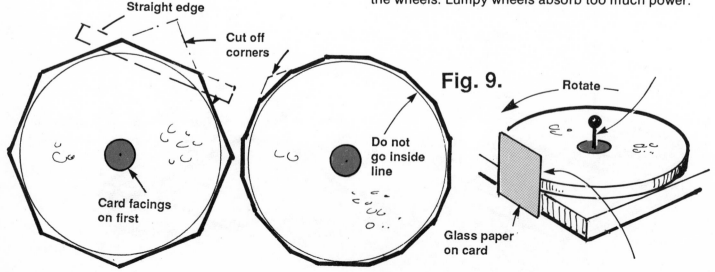

Straight edge

Cut off corners

Card facings on first

Do not go inside line

Fig. 9.

Rotate

Glass paper on card

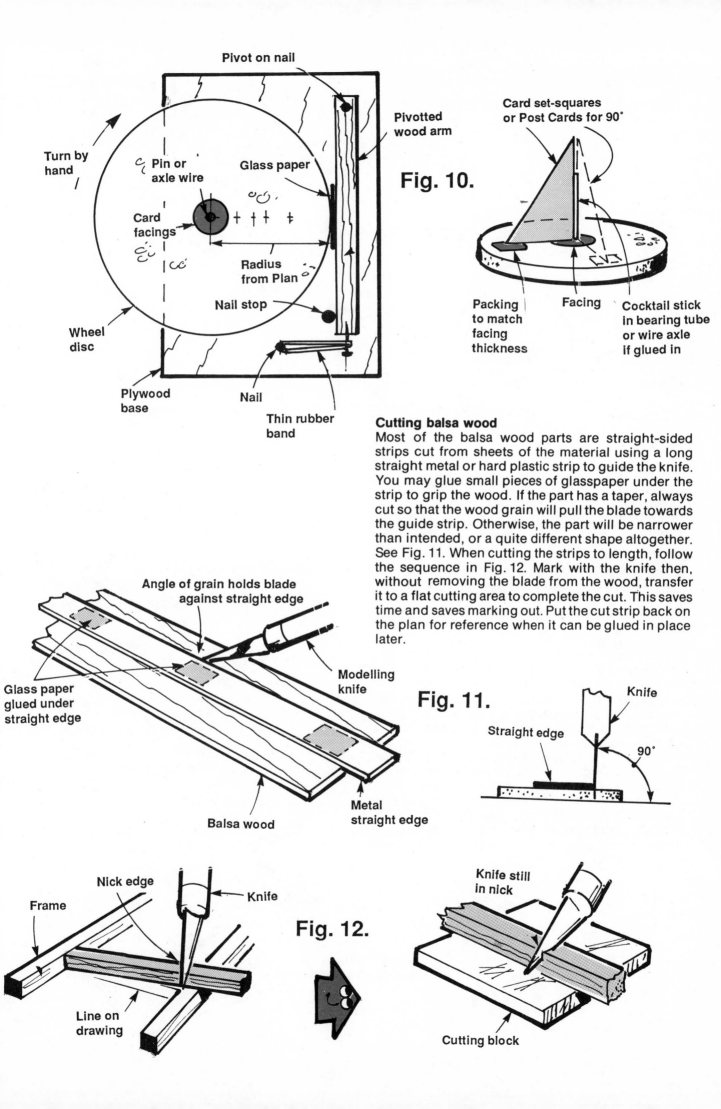

Pivot on nail

Turn by hand

Pin or axle wire

Glass paper

Pivotted wood arm

Fig. 10.

Card facings

Radius from Plan

Nail stop

Wheel disc

Plywood base

Nail

Thin rubber band

Card set-squares or Post Cards for 90°

Packing to match facing thickness

Facing

Cocktail stick in bearing tube or wire axle if glued in

Cutting balsa wood

Most of the balsa wood parts are straight-sided strips cut from sheets of the material using a long straight metal or hard plastic strip to guide the knife. You may glue small pieces of glasspaper under the strip to grip the wood. If the part has a taper, always cut so that the wood grain will pull the blade towards the guide strip. Otherwise, the part will be narrower than intended, or a quite different shape altogether. See Fig. 11. When cutting the strips to length, follow the sequence in Fig. 12. Mark with the knife then, without removing the blade from the wood, transfer it to a flat cutting area to complete the cut. This saves time and saves marking out. Put the cut strip back on the plan for reference when it can be glued in place later.

Angle of grain holds blade against straight edge

Modelling knife

Glass paper glued under straight edge

Balsa wood

Metal straight edge

Fig. 11.

Knife

Straight edge

90°

Nick edge

Knife

Frame

Fig. 12.

Line on drawing

Knife still in nick

Cutting block

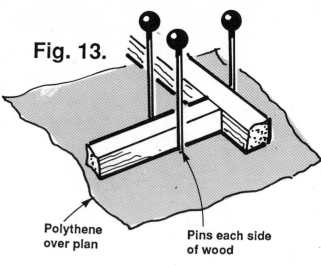

Fig. 13.

Polythene over plan

Pins each side of wood

Fig. 14.

Transfer lengths

All measurements from here

Paper tick strip

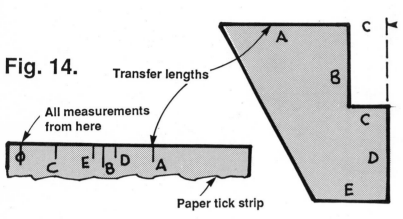

Fig. 15.

Greaseproof paper

Masking tape

Trace part

Prick through

Masking tape

Close spacing on curves

Miss chamfered tile edge

Straight edge

Join up holes with fine felt pen

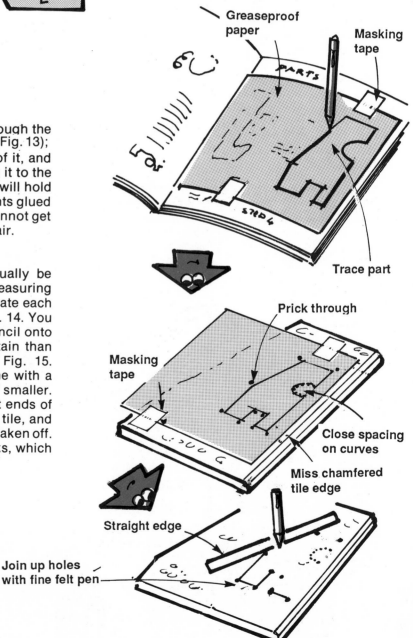

Gluing wood joints

Use pins on each side of the strips, not through the ends where they may split the wood (Fig. 13); polythene between the drawing, or copy of it, and the assembly will prevent the glue sticking it to the plan. If you use epoxy, then masking tape will hold the joint temporarily. Do not do this on joints glued with PVA: it slows the drying because air cannot get to the joint. Epoxy however, sets without air.

Transferring shapes

The shapes shown on the plans can usually be marked out on the tiles, card or balsa by measuring with the compass or a strip of paper. Indicate each line by a letter to avoid mistakes, as in Fig. 14. You can trace more complex shapes with a pencil onto greaseproof paper, which is easier to obtain than tracing paper. Follow the sequence in Fig. 15. Pricking through onto the tile is best done with a cocktail stick, as the holes will close up if smaller. You will also find that you can push short ends of cocktail sticks through the paper into the tile, and they will remain in place when the paper is taken off. You can then draw lines between the sticks, which you can remove and save for future use.

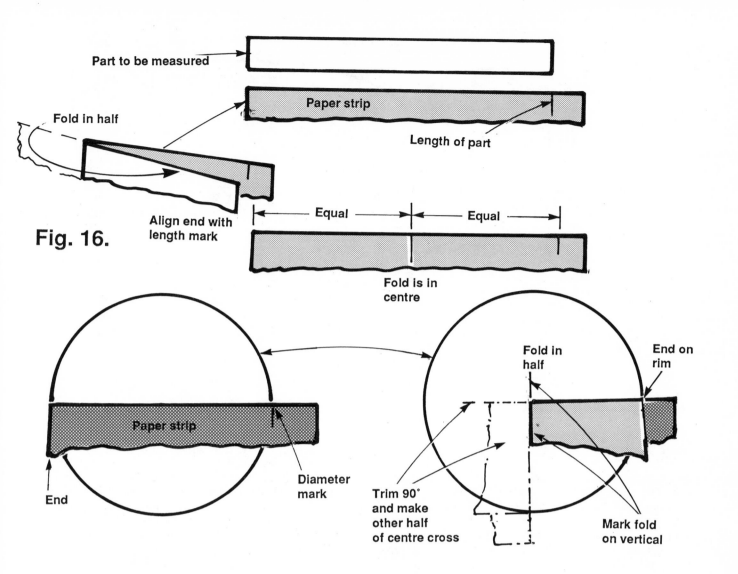

Part to be measured

Paper strip

Length of part

Fold in half

Align end with length mark

Fig. 16.

Equal | Equal

Fold is in centre

Paper strip

End

Diameter mark

Fold in half

End on rim

Trim 90° and make other half of centre cross

Mark fold on vertical

Finding the centre

Take a strip of thin paper and mark the length of the component on one edge. Fold the paper in half, so that the two marks meet. The fold line is the centre – mark it on the component. Try this for round items, such as the ends of containers, which may be used as pulleys or wheels (Fig. 16), or make the centre finder seen in Fig. 17. This is quicker and, if correctly made, more accurate.

Making holes

Most of the materials are soft, so you can start holes with a pin and open them out with a cocktail stick, or by rotating the tip of the modelling knife gently in the pinhole. Luckily, the hole in a sequin is a tight fit on a giant paper clip, so these make excellent washers to retain wheels and cranks on axles.

Plywood 90° "L" piece

Wheel or tube

Mark again

Turn 90°

45°

Diagonal of card in corner of "L" piece

Card 90° 45° triangle

Align this edge

Fig. 17.

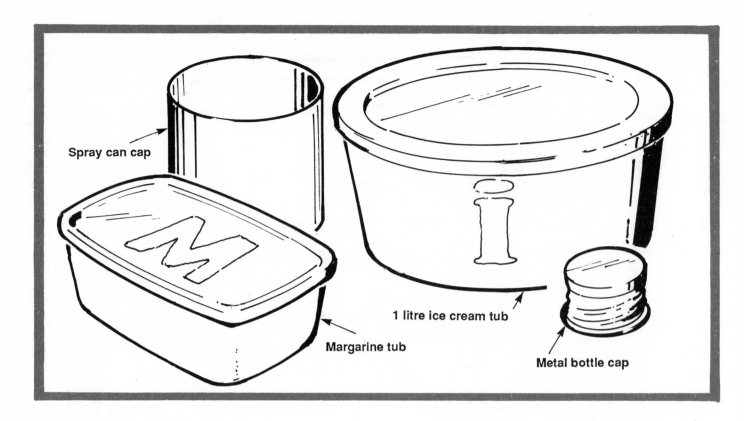

Spray can cap

Margarine tub

1 litre ice cream tub

Metal bottle cap

WHAT DRIVES THE MODELS?

How 'water power' works is varied and interesting. For example, the first model – Chicken – is a simple application of a falling drop of water, which unbalances the system causing it to move. Others use water in larger quantities, automatically fed into the system via a simple valve, and these machines use its imbalance to drive them. Another uses pressure built up through a vertical tube. There is also an orbiting water wheel and a simple water clock...and so on.

These models are a fascinating demonstration of simple mechanical principles. I hope that they will encourage you to design and make other variations on these themes and to develop an even greater interest in the world of modelling and machines. Also, and just as importantly, I hope you find as much pleasure and enjoyment in making the models as I did in designing them.

1 Chicken

IN THIS model, water drops fall into a trough, causing it to tilt, which tips the drops off and allows the trough to rise again. If you fix the trough to the neck of a cut-out 'chicken', the bird will peck until the water is used up.

Step 1 Lay the greaseproof paper over the drawing and trace the shape of the body and neck. Put the tracing over a tile and, with a cocktail stick, prick around the outlines only. Remove the paper and join up the marks with the pen. Use the paper again to mark the position of the pivot holes.

Cut out the shapes with a sharp modelling knife. Cut four square facings of card to reinforce the pivots. Chop off a piece of cotton bud stem to form a bearing.

Tools

- modelling knife • straight edge • scissors
- fine felt pen • ruler

Materials

- 2 tiles • postcard • cotton bud stem
- bendy straw • masking tape • pins
- glass-headed pin • 2 tubs from 35 mm. film cassettes • lid of a rectangular margarine tub • giant paper clip • cocktail stick
- greaseproof paper

Glue with epoxy

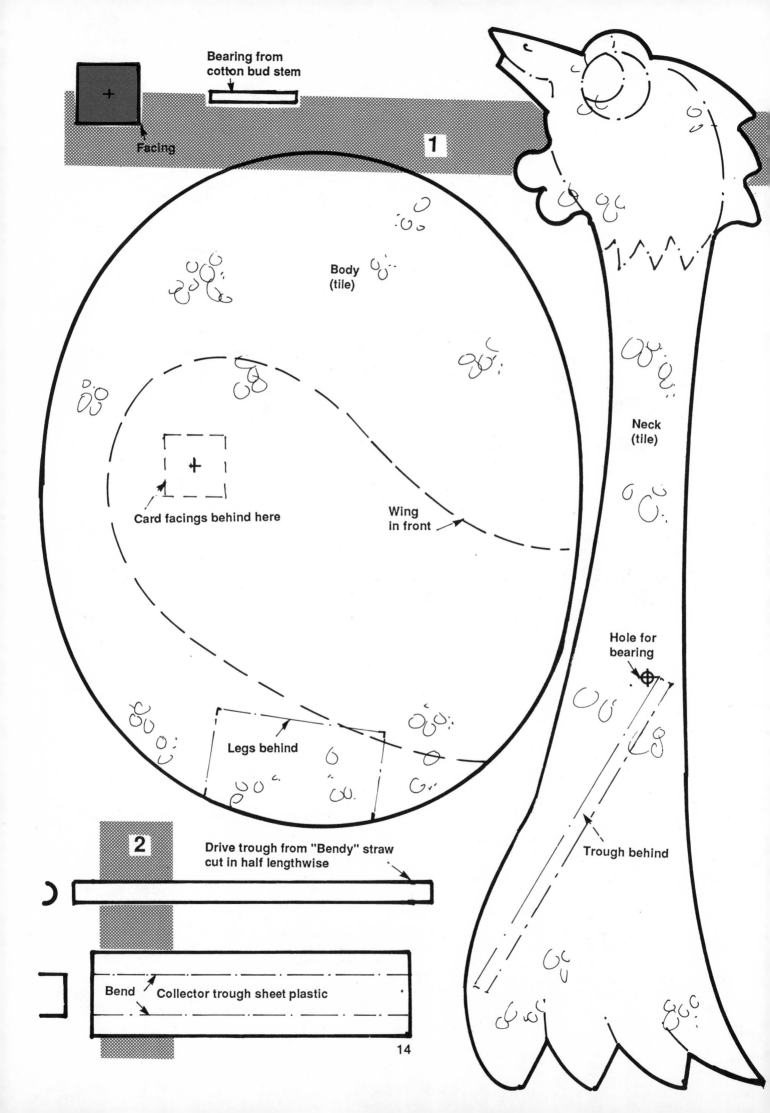

Bearing from
cotton bud stem

Facing

1

Body
(tile)

Card facings behind here

Wing
in front

Neck
(tile)

Hole for
bearing

Legs behind

2

Drive trough from "Bendy" straw
cut in half lengthwise

Bend Collector trough sheet plastic

Trough behind

14

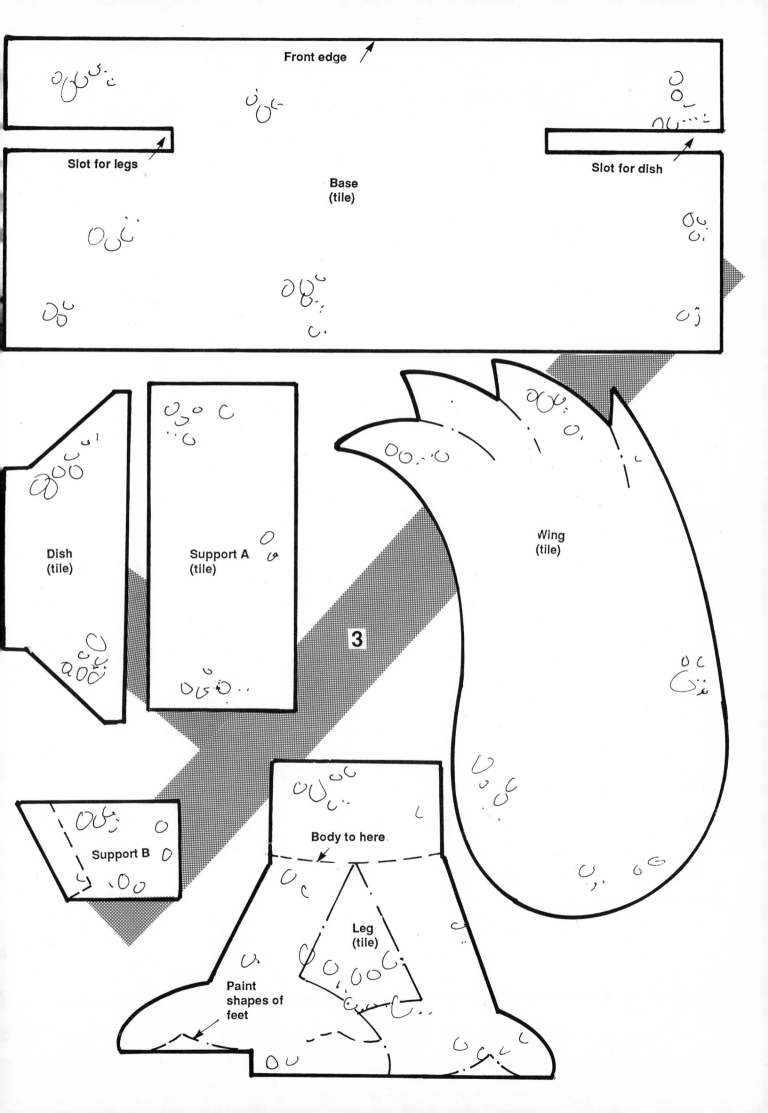

Front edge

Slot for legs

Slot for dish

Base
(tile)

Dish
(tile)

Support A
(tile)

Wing
(tile)

3

Support B

Body to here

Leg
(tile)

Paint
shapes of
feet

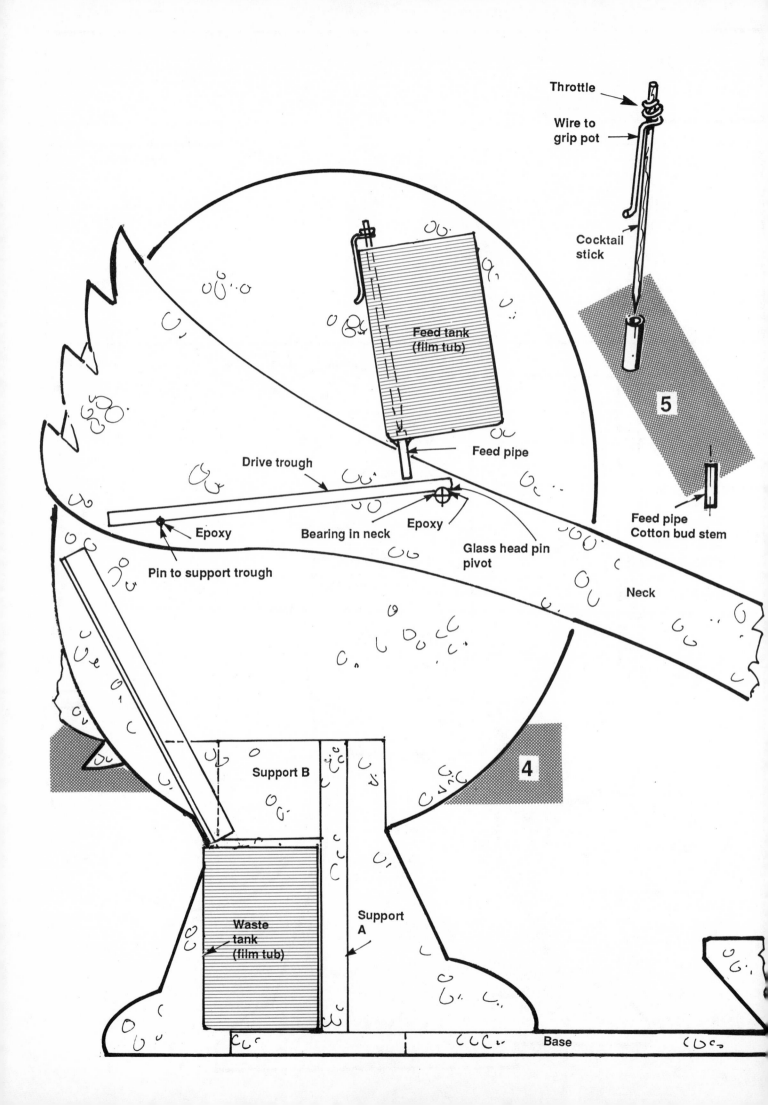

Throttle

Wire to grip pot

Cocktail stick

5

Feed pipe
Cotton bud stem

Feed tank
(film tub)

Feed pipe

Drive trough

Epoxy

Bearing in neck

Epoxy

Pin to support trough

Glass head pin pivot

Neck

4

Support B

Support A

Waste tank
(film tub)

Base

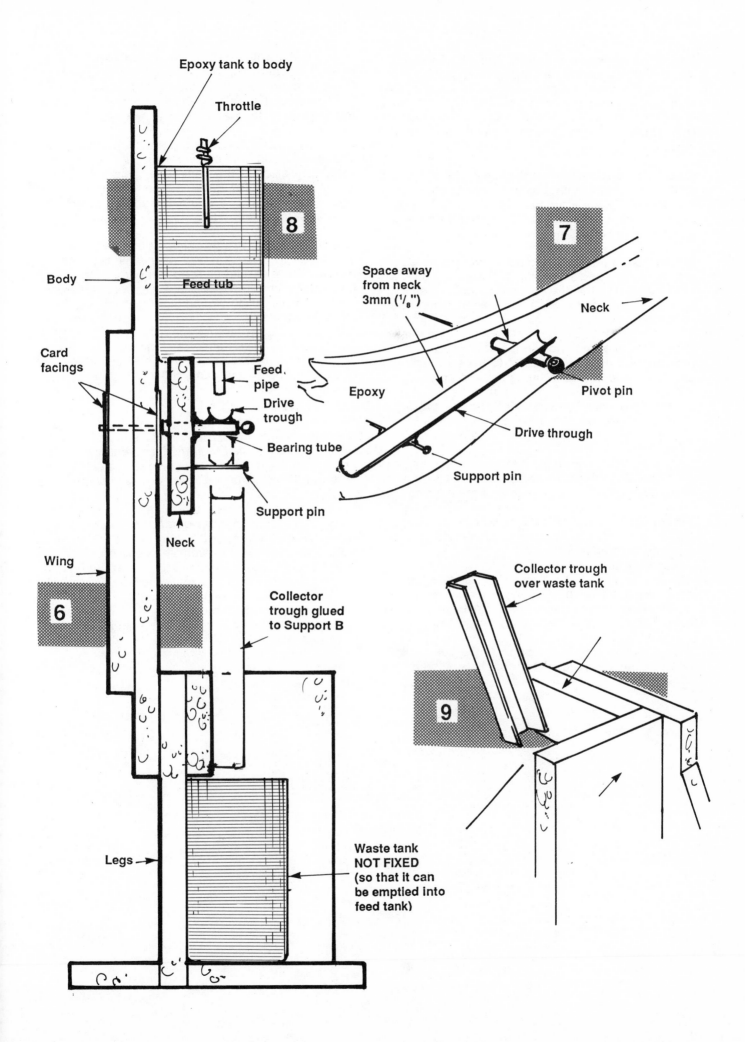

Epoxy tank to body

Throttle

8

Body

Feed tub

Card
facings

Feed
pipe

Drive
trough

Bearing tube

Support pin

Neck

Wing

6

Collector
trough glued
to Support B

Legs

Waste tank
NOT FIXED
(so that it can
be emptied into
feed tank)

7

Space away
from neck
3mm ($^1/_8$")

Neck

Epoxy

Pivot pin

Drive through

Support pin

Collector trough
over waste tank

9

Step 2 Cut a piece of the straw, then carefully slice it lengthwise, to form a half-round drive trough.

Using the ruler and pen, mark out a piece of thin sheet plastic from the lid of a margarine tub. Crease it with the ruler and fold the sides up to make a rectangular collector trough, which will catch the splashes.

Step 3 Trace and transfer the shapes of the legs and wing onto another tile. Using a ruler or paper strip for measuring, mark out the two supports, base and dish. Cut all these parts out. The straight edge will guide the knife on the last four pieces.

Step 4 Using fast epoxy, glue the wing and legs to the body. Pins and small pieces of masking tape will hold the joints while they set. Continue by gluing the legs and dish into the base, and the supports to the rear of the legs.

Step 5 Make a small hole in the bottom edge of one film tub, or a small spice tub, to form a feed tank. Cut another piece of cotton bud stem and epoxy it into the hole, facing down, to form a feed pipe. Twist a paper clip round a cocktail stick and bend it to grip the edge of the tub. The bottom point of the stick will go into the tube. This forms a throttle valve.

Step 6 Epoxy the card facings, one on each side of the body/wing at the pivot point, and each side of the neck. Temporarily plug one end of the bearing tube (Step 1) and epoxy it into the neck projecting $1/8''$ (3mm) at the front, which faces the body. Remove the plug, which should have prevented it being clogged with epoxy.

Push the glass-headed pin through the tube and into the body, allowing the neck to rock up and down. With pliers, snip off the point of the pin where it comes out of the wing. Do this with the point facing away from you as it will fly off.

Step 7 Epoxy the drive trough to the bearing tube and to a pin in the neck. It should be spaced at least $1/8''$ (3mm) away from the neck and parallel to it. Make it sit there without rocking over by putting a scrap of masking tape across until it sets.

The rear end of the trough tilts down slightly when the beak is on the base. Refer to Step 4.

Step 8 Epoxy the feed tank to the back of the body. The feed pipe should be over the trough and the pot should be clear of the neck when down or up just past horizontal. Use masking tape to hold it while it sets. Refer to Step 4 again for angle, and to the section in this step.

Step 9 Epoxy the plastic collector trough to support 'B'. It has to clear the end of the neck and drive trough. There should be room beneath for the other film tub, which is the waste tank. Do not fix this tank; it has to be removed when emptying the water back into the feed tank.

If you decide to colour the model, use emulsion paint, tinted with poster colour if necessary. To save weight avoid thick paint on the neck. Just paint the head with poster paint and allow to dry.

HOW IT WORKS

Put a small piece of modelling clay on the rear end of the neck, beside the tail feathers. It should bring the beak just off the base. Fill the feed tank with water, and adjust the throttle by pushing the cocktail stick down until the water dribbles slowly into the drive trough.

The neck should rise and fall again as the water collects in the trough, then gets tipped into the waste tank. Adjust the amount of clay to get the balance right. Too much will prevent the beak reaching the base, too little will stop it rising. Adjust the throttle so that the movement is even, or erratic if you wish. Too much water merely gets wasted.

Like most of the models in this book, some splashes of water are inevitable, so operate it out of doors or where the odd splash can be wiped up easily – the kitchen for example. Otherwise put it on a plastic or metal tray.

2 Aqua Jet

THIS MODEL demonstrates the power of water pressure.

Raise a tank of water well above the end of a pipe. The weight of water produces pressure. A jet of water from a short pipe will not be as powerful as one from a long one. That jet is going to do some work. It points at a set of corrugations, and squirts itself backwards. Imagine a turbine – a wheel with blades on the edge, in the stream of a fixed jet which makes the wheel spin – the Aqua Jet shows that the idea works backwards. Fix the blades and let the jet spin. The blades are a strip of corrugated cardboard around a waste tank, and the pipe that builds up the pressure is made from straws, which also serve as a drive shaft. There are no watertight bearings or other complicated features, so it is a good introductory model to build.

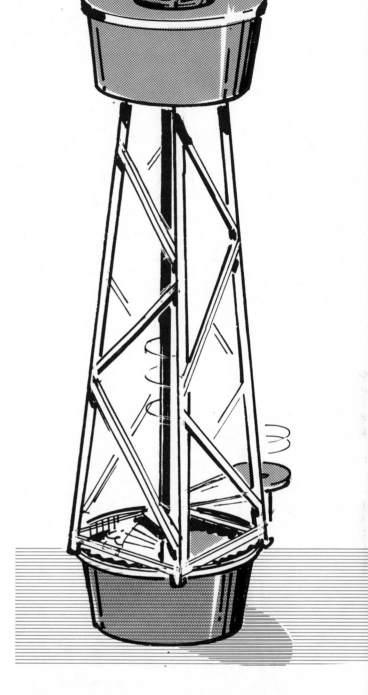

Tools

- modelling knife • straight edge • scissors
- ruler • pliers • compass • fine felt pen

Materials

- ⅛″ (3mm) balsa sheet • corrugated cardboard • 3 1-litre size round ice cream tubs
- 2 bendy straws • a thin straw – about ⅛″ (3mm) diameter • 3 giant paper clips
- a cotton bud stem • sewing thread
- a cocktail stick • candle wax • masking tape • greaseproof paper

Glue with epoxy

Step 1 Using the straight edge to guide the knife, cut strips of balsa wood. Five ¼″ (6mm) wide strips, each the length of the page from top to bottom, will be required. Cut 12 more, each ⅛″ (3mm) wide, which makes them square-sectioned.

Trace the tower side onto the greaseproof paper, and tape that onto a couple of layers of cardboard carton to form a building surface. Rub the wax onto the position of each joint, so that the glue will not stick the wood to the tracing.

Assemble the four tower sides on the tracing, using epoxy glue. All the joints are overlapped for ease of building. The strips can be lapped past each other, then trimmed back afterwards. Note that one edge of each tower side is narrower than the other.

Step 2 When all four are set, glue each narrow side behind the wide side of the next. Thus there will be four 'L'-section uprights and all the bracing and cross struts on the outside. Use masking tape to hold the corners together while they set.

The finished tower should sit squarely on the rim of one of the ice cream tubs. This will be the waste tank and has no lid.

Step 3 With the compass, mark out a circular top piece for the tower. Cut this out, using short strokes, with the modelling knife, and pierce the centre for the thin straw. Do not fix it yet to the tower.

Step 4 With pliers, form a paper clip into an axle. Note that it is offset to fit onto a bendy straw. Smooth the bottom end to a slightly rounded shape by rubbing on glasspaper or on a paving stone. Bind it to the straw as shown. Form another paper clip into a

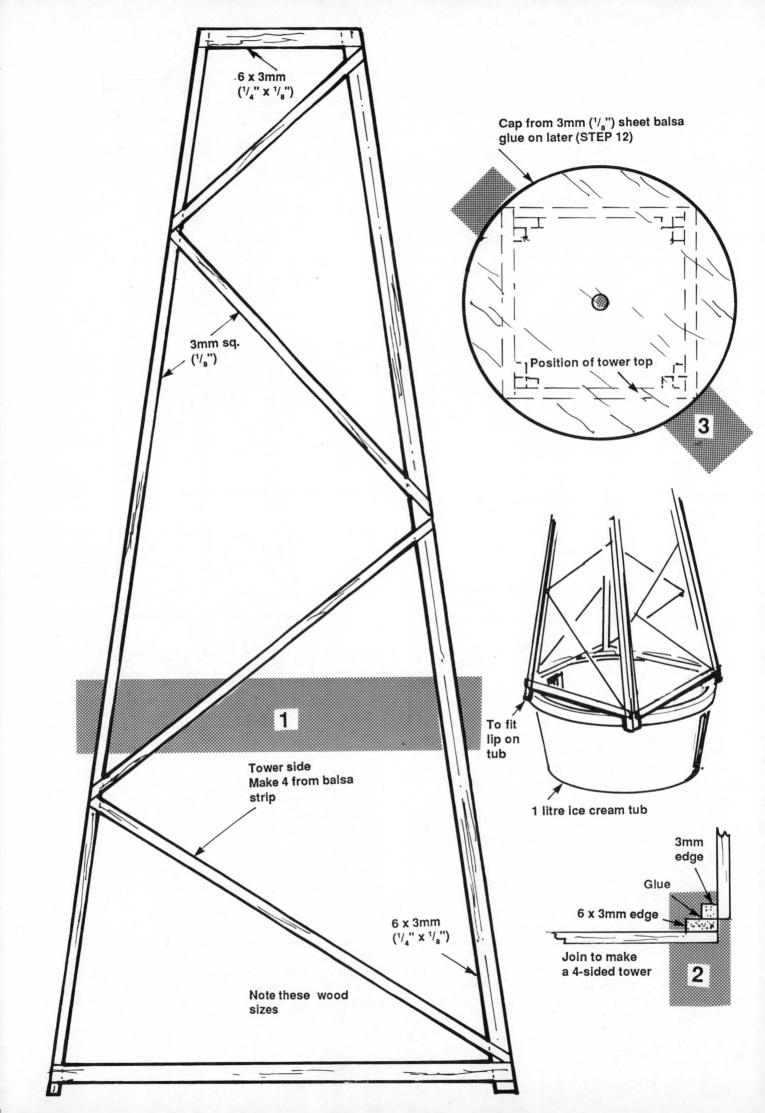

.6 x 3mm
(¹/₄" x ¹/₈")

3mm sq.
(¹/₈")

1

Tower side
Make 4 from balsa
strip

6 x 3mm
(¹/₄" x ¹/₈")

Note these wood
sizes

Cap from 3mm (¹/₈") sheet balsa
glue on later (STEP 12)

Position of tower top

3

To fit
lip on
tub

1 litre ice cream tub

3mm
edge

Glue

6 x 3mm edge

Join to make
a 4-sided tower

2

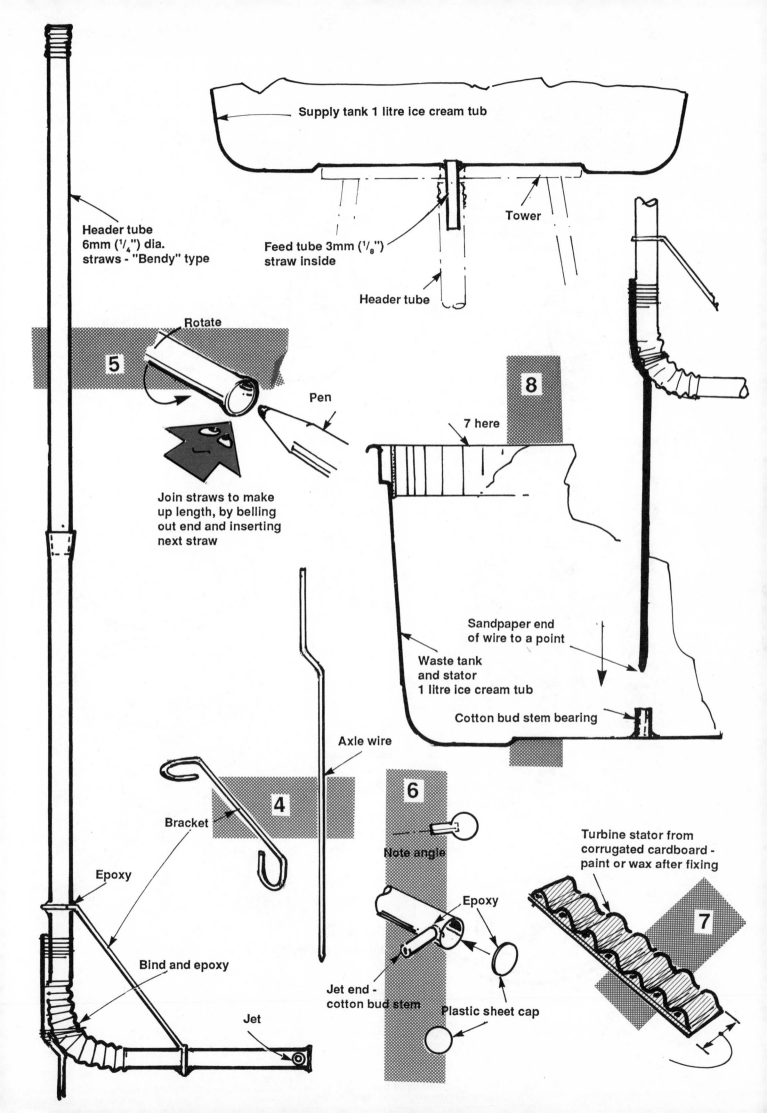

Header tube
6mm (1/4") dia.
straws - "Bendy" type

Supply tank 1 litre ice cream tub

Feed tube 3mm (1/8")
straw inside

Tower

Header tube

5

Rotate

Pen

Join straws to make
up length, by belling
out end and inserting
next straw

8

7 here

Sandpaper end
of wire to a point

Waste tank
and stator
1 litre ice cream tub

Cotton bud stem bearing

Axle wire

4

Bracket

Epoxy

Bind and epoxy

Jet

6

Note angle

Jet end -
cotton bud stem

Epoxy

Plastic sheet cap

Turbine stator from
corrugated cardboard -
paint or wax after fixing

7

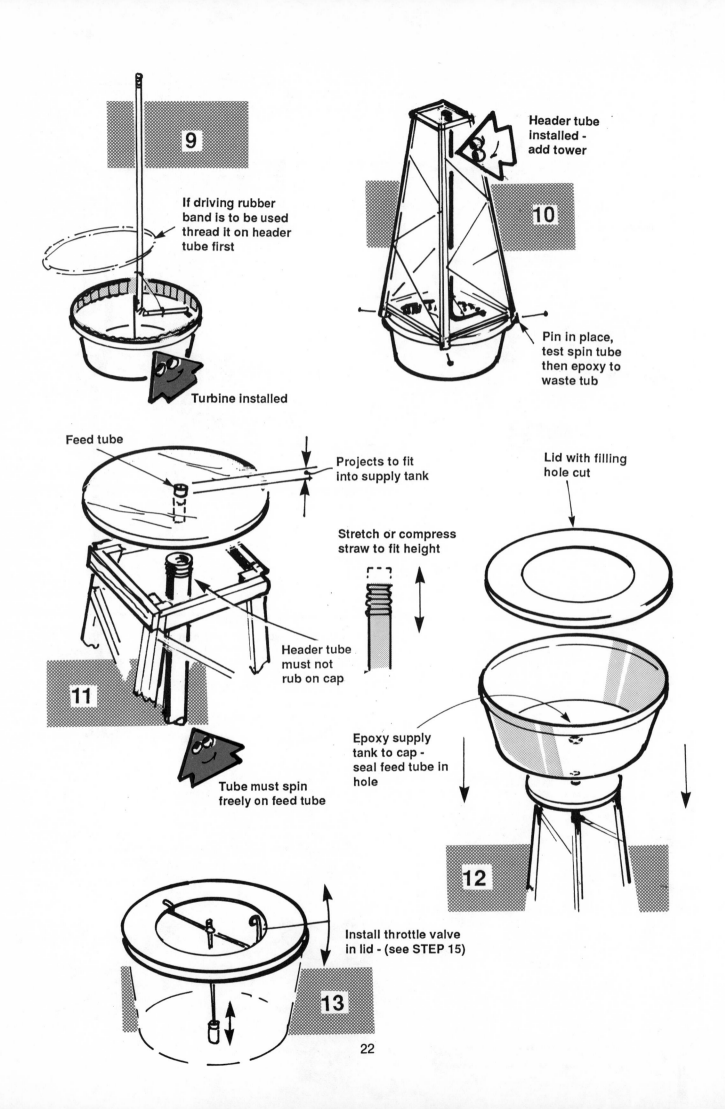

9

If driving rubber band is to be used thread it on header tube first

Turbine installed

Header tube installed - add tower

10

Pin in place, test spin tube then epoxy to waste tub

Feed tube

Projects to fit into supply tank

Stretch or compress straw to fit height

Lid with filling hole cut

Header tube must not rub on cap

11

Tube must spin freely on feed tube

Epoxy supply tank to cap - seal feed tube in hole

12

Install throttle valve in lid - (see STEP 15)

13

22

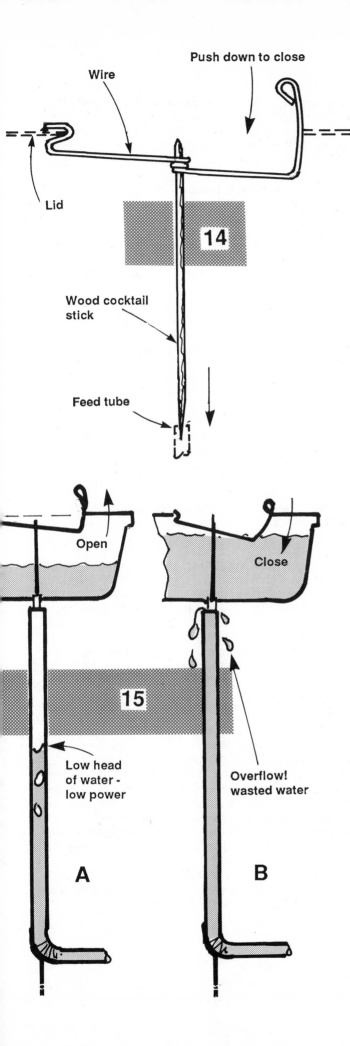

Push down to close

Wire

Lid

14

Wood cocktail stick

Feed tube

Open

Close

15

Low head of water - low power

Overflow! wasted water

A

B

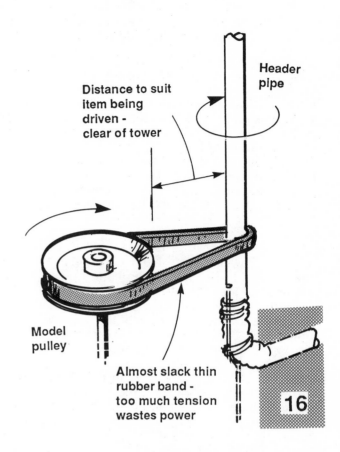

Distance to suit item being driven - clear of tower

Header pipe

Model pulley

Almost slack thin rubber band - too much tension wastes power

16

bracket to go around the straw, holding it to a right angle. Epoxy this and the binding to the straw.

Step 5 Insert a pencil into the top of the straw to bell it out to fit over another one. This forms the 'header tube' which builds up the pressure. Insert the second straw.

Step 6 At the side of the lower bent end of the header straw, make a small hole to take a piece of cotton bud stem. This forms the jet. Epoxy it in without clogging it. Cut a scrap of plastic sheet from the discarded tub lid to form a disk to block the end of the bendy straw. Epoxy it in place. Now the water can only come out of the jet.

Step 7 The turbine stator is a strip of single-sided corrugated cardboard. If none can be found, dampen one side of double-faced corrugated card-board, used for large cartons, leave it for a while, then peel the wet surface away leaving the corrugations exposed.

Cut a ½" (12mm) wide strip across the corrugations. Epoxy it inside the rim of the waste tank. Waterproof it by melting candle wax over it.

Step 8 Cut a short piece of cotton bud stem and epoxy it to the exact centre of the inside of the waste tank bottom. The best way to do this without clogging it is to place it in dry, then put epoxy all round it. Do not lift it or the epoxy will run in. This forms a bearing socket for the axle.

Step 9 Insert the axle into its bearing in the waste tank. Should you decide to drive anything from this model, put a long thin rubber band on the header tube. This will avoid having to take the model apart to fit one later. Hold a spare piece of small straw above, to go loosely into the top of the header tube, to hold it upright.

Test spin the latter. It should not wobble at the lower end, and the jet should not rub on the turbine. Adjustment can be made by gently bending the axle, but not so much as to make the tube run unevenly at the bottom end.

Step 10 If you wish to paint the tower, do it now. Place the tower over the header and set its feet over the edge of the waste tank. Put short pins through the ends of the legs into the rim, without loosening the turbine. Give a final spin to the tube to make sure that the jet clears the bottom of the tower. If it does not, gently bend the end of the straw down. Now epoxy the tower to the tank top.

Step 11 Make a central hole in the bottom of the next ice cream tub, which is the supply tank. Epoxy the feed pipe into the balsa disk, so that it projects into the supply tank. Place the disk on top of the tower, so that the feed tube enters the header tube freely. The header must not rub on the balsa. If it does, trim it slightly. If it is too short to fit over the feed tube, stretch it by extending the bendy bellows part near the top.

The clearance between the top of the header and the disk must not be so great as to allow the axle to come out of the bearing. Stretching the bellows corrects this.

Epoxy the disk centrally on the tower with the short end of the feed tube uppermost and the lower end in the header.

Step 12 Cut a central hole in the lid of the supply tank. The size is not important: about half its rim diameter will do. Epoxy the supply tank to the disk, with the feet tube in the hole. Seal around the tube with epoxy. Put a tapered piece of modelling clay or wax in the end to avoid clogging the tube. This plug should not be so small as to drop through into the header as it might obstruct the flow later.

Step 13 Make a throttle valve by bending a paper clip around a cocktail stick. The shape of the ends will depend on the size of the hole in the lid, but the bottom point of the stick should enter the straw. One end of the clip hooks over the edge of the hole, while the other rubs on the opposite edge.

Step 14 This illustration shows the correct positioning of the throttle. The point of the stick will not seal the tube, but will merely slow the flow of water. Notice how the handle end is bent to an arc, so that it holds its position by friction.

Step 15 Pass the rubber band out through each side of the tower and secure it clear of the tube with masking tape.

Test run the model, as follows. Close the throttle and fill the supply tank, via the hole in its lid, using the third ice cream tub to carry the water. After spluttering over some bubbles, the jet should drive the water against the turbine stator and start turning slowly. Look at the header tube. You should be able to see the water level in it. If it appears as at 'A', the head of the water is low because the feed is not fast enough. A low head of water lacks power.

Open the throttle, by lifting the clip handle, and the head of water will rise and the speed increase. Should water spill out of the top of the header, as in 'B', then the throttle is open too wide and the supply is too fast for the jet to cope with, even at the now higher pressure. Reduce the throttle setting until the header is full but not overflowing.

As the water level in the supply tank falls, the speed will drop. Now you can open the throttle to compensate, because the pressure at the throttle valve is less so the head gets lower.

You might think that the small size of the jet would prevent the water being used up so fast. The higher pressure, produced by the height of the top of the header above the jet, ensures that more water goes through the jet than would be the case if the supply was at the same height as the jet.

When the water is used up, tilt the model so that it can be emptied from the waste tank into the spare tub, then, from that, back into the supply tank. The remaining part of the lid on this tank should stop any last drops of water spilling when the model is tilted.

Step 16 To drive another model, the rubber band should pass through the tower and go around the driven pulley on that model. Unless high speed, low power is needed, it would be wise to choose the pulley so that it is at least four times the diameter of the header pipe. The sketch shows the pulley with the shaft vertical, although it may be horizontal with the belt twisted but not completely crossed (which produces friction).

The tension of the belt has to be very light. Only enough is required to stop it slipping on the header. Also the belt should not rub on the tower.

3 Mystic Fountain

NOW FOR a party trick! A small tank sits on a bracket above a larger tank, which is partly filled with water. There is a little water in the top tank and a spout that is plugged. Remove the plug. Perhaps a little water will dribble out, then stop. You would expect that.

What actually happens is that the water flows out of that spout, and goes on flowing, with increasing force! It puts more and more into the lower tank than could possibly come out of the little top tank. There is no outside connection for the water, no motors whirring.

Eventually, before the bottom tank overflows, the top tank stops pouring. Is this a reversal of the 'head of water' system used in the previous model? Make the fountain and find out.

Tools

- modelling knife   - straight edge   - ruler
- fine tip felt pen

Materials

- 2 tiles   - a 500g size rectangular margarine tub   - a transparent film tub, or similar small-sized clear container   - a bendy straw
- a cotton bud stem   - pins   - a cocktail stick   - a small piece of thick balsa wood or several scraps glued together   - a round toy balloon of average size

Glue with epoxy

Pressure pad
4 off

1

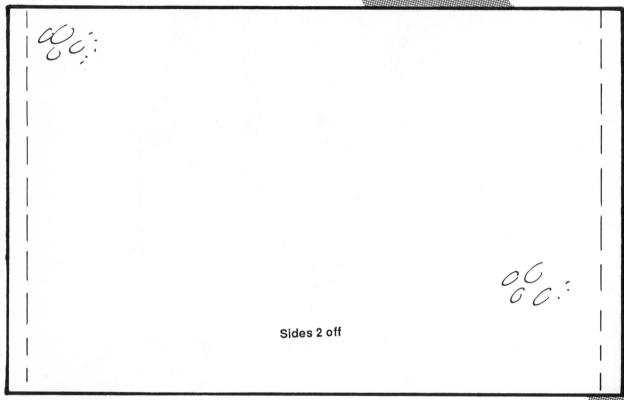

Sides 2 off

Box end
B

Box end
A

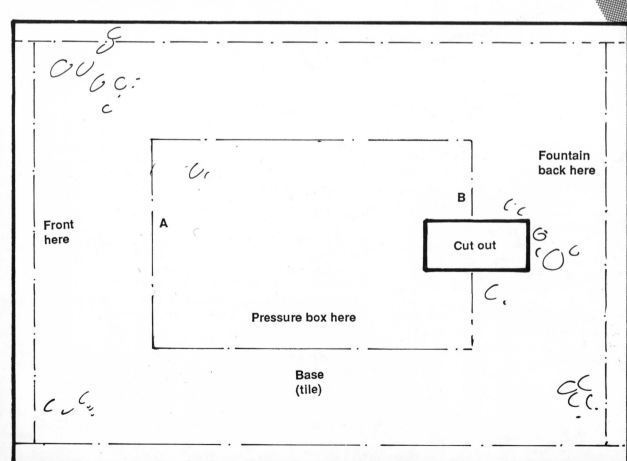

Front
here

Fountain
back here

B

A

Cut out

Pressure box here

Base
(tile)

Pressure
box side
2 off

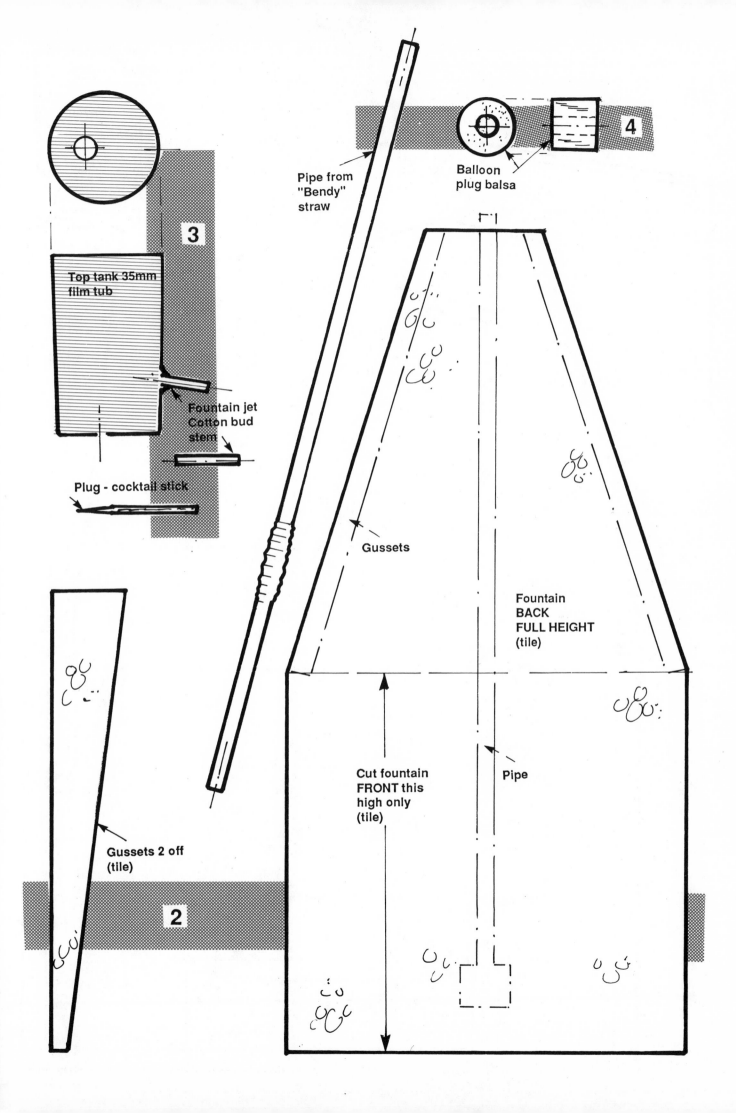

3

Top tank 35mm film tub

Fountain jet Cotton bud stem

Plug - cocktail stick

Pipe from "Bendy" straw

Balloon plug balsa

4

Gussets

Fountain
BACK
FULL HEIGHT
(tile)

Pipe

Cut fountain
FRONT this
high only
(tile)

Gussets 2 off
(tile)

2

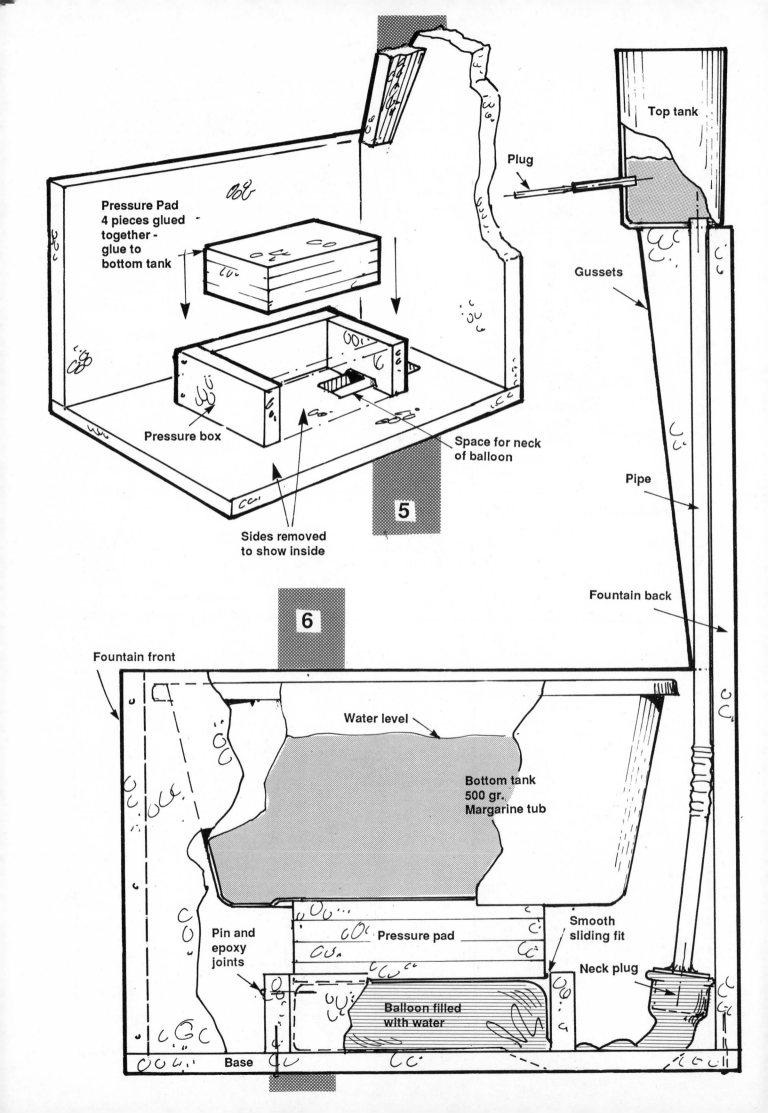

Pressure Pad 4 pieces glued together - glue to bottom tank

Pressure box

Sides removed to show inside

Space for neck of balloon

5

Top tank

Plug

Gussets

Pipe

Fountain back

6

Fountain front

Water level

Bottom tank 500 gr. Margarine tub

Pin and epoxy joints

Pressure pad

Smooth sliding fit

Neck plug

Balloon filled with water

Base

Step 1 Measure out the parts for the fountain base and pressure box onto a tile. Cut these out with a modelling knife, and don't forget the hole and notch shown on the drawing. Make four pressure pad pieces, and two each fountain sides and pressure box sides.

Step 2 In the same way, produce the fountain back and two brackets. Note that the front end of the fountain is shown on the drawing of the fountain back.

Step 3 Make a hole in the bottom, off centre to the rear. It should be large enough to take the bendy straw end. Start the hole with a pin, then rotate the tip of the modelling knife in it, using only light pressure to enlarge it.

Make another hole some 1/4" (6mm) up from the bottom of the front face. It should be large enough to take a short piece of cotton bud stem, which you then epoxy in, sloping down slightly, to form the fountain jet.

Cut a short piece from one end of a cocktail stick to form a plug to fit in the end of the jet.

Step 4 Cut a piece of balsa wood to form a round plug to fit tightly into the neck of the balloon. Make a hole through the centre to fit the end of the straw. The plug should be smooth and free from cracks and notches.

Balsa wood is not completely airtight or watertight. Smear it all over with epoxy to seal it and to fix it to the short end of the bendy straw.

Step 5 Epoxy the four pressure pads pieces together to make a solid block. See that the glue does not ooze out as blobs at the edges will stop it fitting into the box.

Assemble the pressure box on the fountain base, using epoxy and short pins to reinforce the corners. The pins should not appear inside the box. Test the clearance with the pressure pad. It should slide freely right down into the box.

Assemble the fountain sides, front and back to the base, and add the brackets to support the top tank in an upright position.

Step 6 Place the pressure pad in the pressure box with a smaller scrap of tile under it. Check that the margarine tub is free from grease outside and in.

Put a few spots of epoxy on the top surface of the pressure pad and press the margarine tub onto it. There should be clearance all round the tub. Space it equally from the sides with folded paper, until set.

Mark one end to enable it to be fitted in the right way round, then lift it out. Stick narrow strips of plastic, cut from the discarded lid, inside the sides and front of the base. These will help the tub to slide smoothly.

Blow up the balloon, leave it for half an hour, then let it down. It will be crumpled, but larger than before

it was blown up. Remove the tile packing from the pressure box and put the deflated balloon in it. The neck goes through the hole in the base and back up between the outside of the pressure box and the inside of the fountain back.

Push the plug into the neck of the balloon. Bind the neck with thread to make a good seal. Push the balloon down through the hole in the base, under the end of the pressure box and up onto it. The neck will still be outside the box and the straw facing upwards. Remove any large folds in the balloon and neck, and squeeze as much air out as you can.

Epoxy the top of the straw to the top tank and the fountain back. Epoxy the top tank to the gussets, with the jet facing out. Replace the bottom tank, so that the pressure pad rests on the balloon. This completes the construction.

Setting it up
You have to set the model up and, once you have done this, the repeated demonstrations are easy.

Plug the jet
Slowly pour water into the top tank. It should start by first rising, then emptying via the straw. Continue filling. The empty bottom tank should soon be rising, as water enters the balloon. If it does not, there is an air lock.

To cure this, remove the bottom tank and gently squeeze the balloon until bubbles stop coming into the top tank. Continue pouring water into the top tank until the balloon has filled enough to be level with the top of its box. Replace the bottom tank, making sure that the pressure pad is inside the edge of the pressure box.

Remove the plug from the jet. The water will run down to the jet level and stop.

Pour water into the bottom tank until it is almost three quarters full. By this time, the jet should start pouring again. Stop filling the bottom tank at this point. Replace the plug in the jet. Mark the position of the top edge of the bottom tank on the inside of the fountain sides.

All is now ready for the demonstration. Just remove the plug from the jet and watch. To reset the system for the next demonstration, do the following.

Re-fit the plug
Using a small paper cup or similar container, take half a cupful of water out of the bottom tank and slowly pour it into the top tank. As it goes in, the bottom tank will rise. Watch the marks on the fountain sides.

When the tank reaches them, water in the top tank should be at the jet level. Return the rest of the water to the bottom tank. The water in the top tank may rise very slightly.

All is now ready for another demonstration. After a few tries, you will be able to judge how much water has to be transferred.

HOW IT WORKS

The head of water in the straw has to be balanced, because it produces pressure. This is done by the weight of water in the bottom tank. But the bottom tank is lower than the top tank. How can it produce pressure?

There is much more water in the tank than in the balloon but, if you put the balloon in the bottom tank, the water inside it will be at the same pressure as the bottom tank.

If the balloon is under the tank, there will not be much difference because the water will try to flow out sideways. Contain the balloon in a small box.

That box is smaller in area than the tank.

A pressure pad acts like a piston in a cylinder, and the balloon stops the water leaking out around it. Thus, the weight of the water in the tank overcomes the pressure in the balloon, which was caused by the head of water in the straw and top tank. The system stops working when the balloon is flat.

You can put water back into the balloon via the top tank, only by making the bottom tank lighter, i.e. removing some water.

If you want to know what to call it, it's a differential fluid link'!

4 Wacky Well

IN THIS model, a comical character winds the well handle frantically, first one way, then the other, but the bucket he winds only takes water down into the well. In true 'upside-down, backwards and sideways' tradition, it is the bucketfuls of water that make it crank.

The crank has a pulley on it. A piece of thread goes over it, attached to a small bucket at one end and a weight at the other. The weight is heavier than the empty bucket, but a full bucket is heavier then the weight. The bucket fills and empties itself automatically.

Step 1 Using a ruler, mark out the ten shapes of the base, back, well side and foot bracket onto a tile. Cut these out with the modelling knife guided by the straight edge.

Step 2 Mark and cut the side and screen pieces from another tile.

Tools

● modelling knife ● scissors ● straight edge
● pliers ● compass ● fine tip felt pen
● ruler

Materials

● 2 tiles ● corrugated cardboard ● a bendy straw ● a thin straw ● a cotton bud stem
● 2 rectangular 8oz (250g) margarine tubs
● a yoghurt pot ● a giant paper clip
● masking tape ● modelling clay
● a glass-headed pin ● household pins
● 'invisible' nylon thread ● a scrap of balsa wood ● greaseproof paper

Glue with fast epoxy

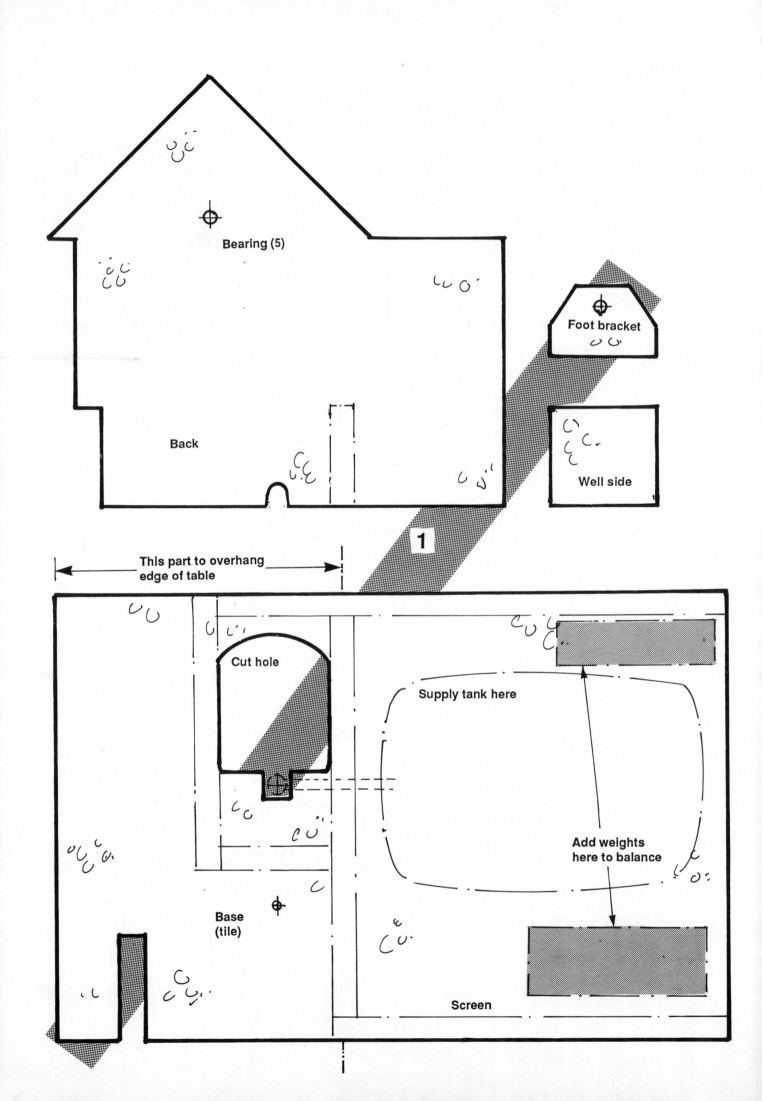

Bearing (5)

Foot bracket

Back

Well side

1

This part to overhang edge of table

Cut hole

Supply tank here

Add weights here to balance

Base (tile)

Screen

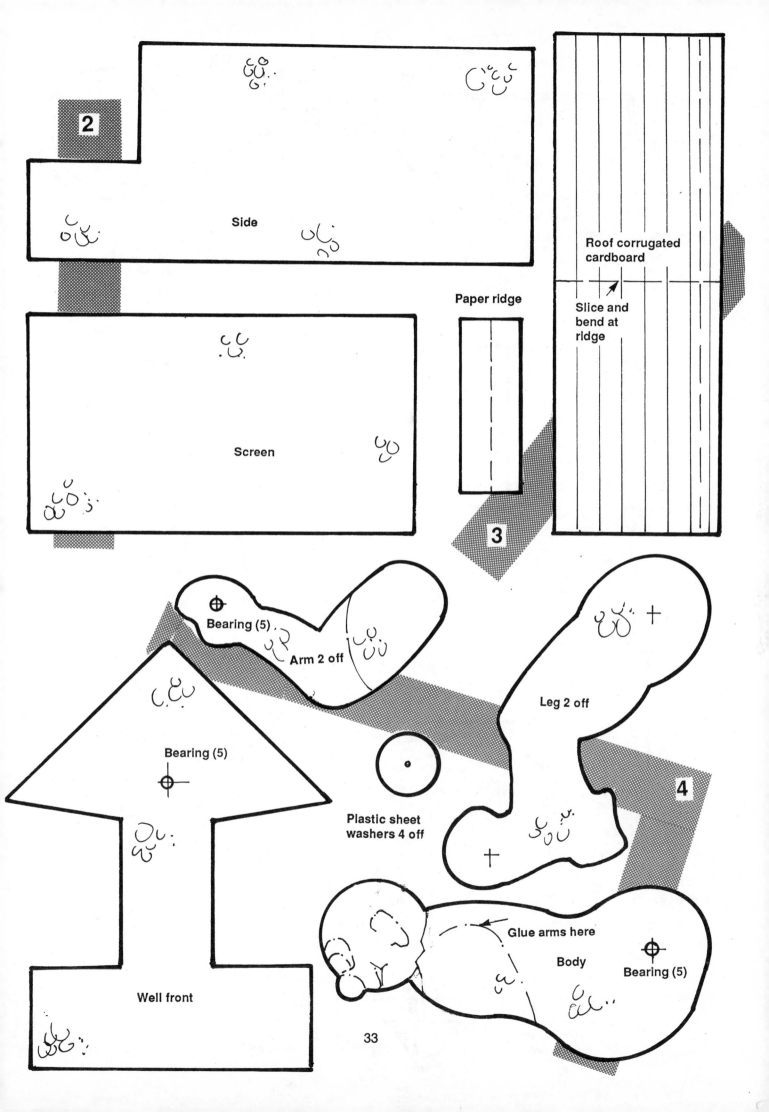

2

Side

Screen

Paper ridge

Roof corrugated cardboard

Slice and bend at ridge

3

Bearing (5)

Arm 2 off

Leg 2 off

Bearing (5)

Plastic sheet washers 4 off

4

Well front

Glue arms here

Body

Bearing (5)

33

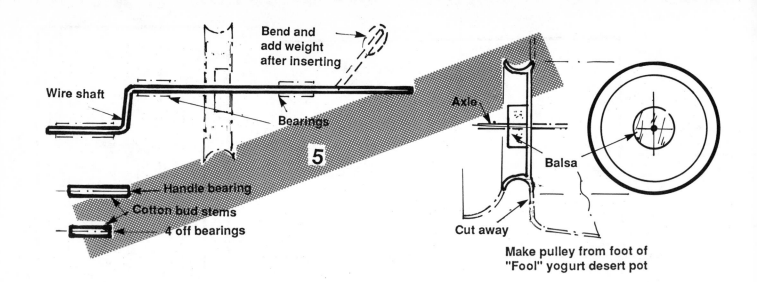

Bend and add weight after inserting

Wire shaft

Bearings

5

Axle

Balsa

Cut away

Handle bearing

Cotton bud stems

4 off bearings

Make pulley from foot of "Fool" yogurt desert pot

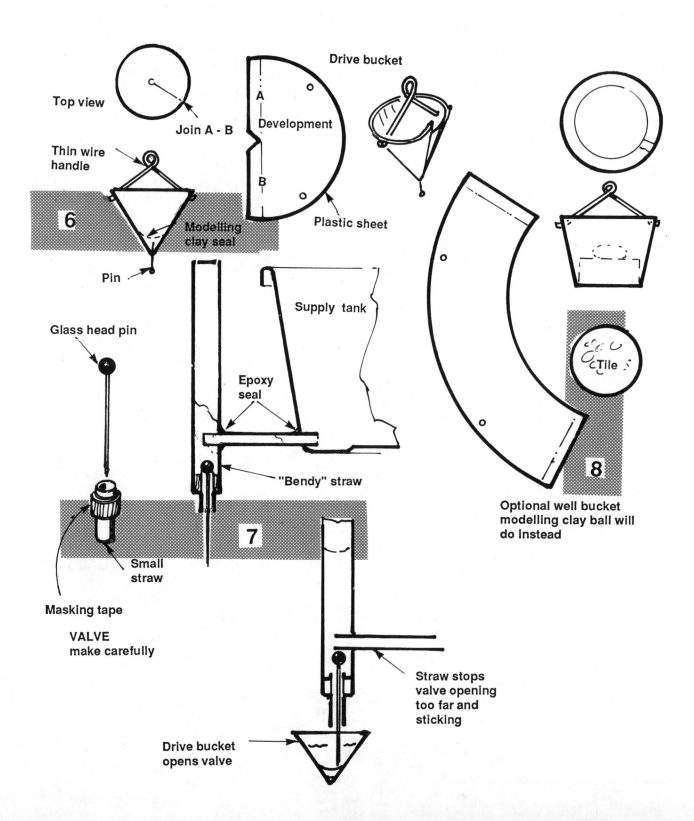

Top view

Join A - B

Drive bucket

Thin wire handle

A

Development

B

6

Modelling clay seal

Pin

Plastic sheet

Glass head pin

Supply tank

Epoxy seal

Tile

"Bendy" straw

7

Small straw

8

Optional well bucket modelling clay ball will do instead

Masking tape

VALVE make carefully

Straw stops valve opening too far and sticking

Drive bucket opens valve

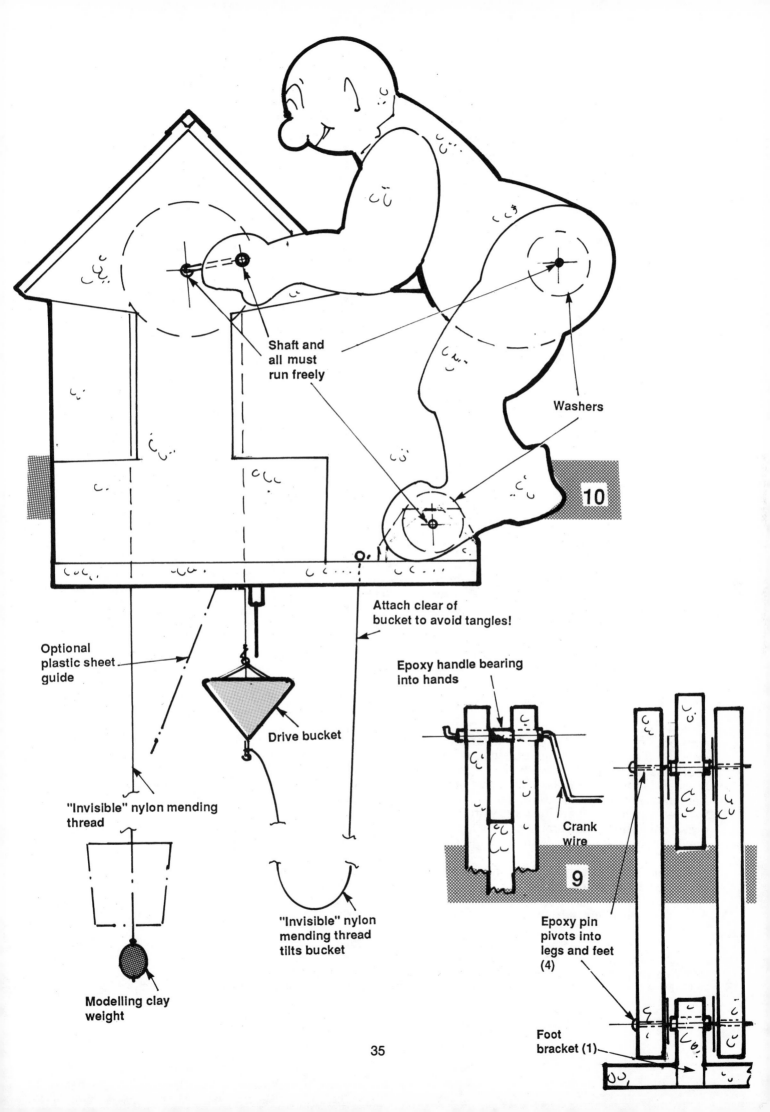

Shaft and all must run freely

Washers

10

Attach clear of bucket to avoid tangles!

Optional plastic sheet guide

Drive bucket

Epoxy handle bearing into hands

Crank wire

9

"Invisible" nylon mending thread

"Invisible" nylon mending thread tilts bucket

Epoxy pin pivots into legs and feet (4)

Modelling clay weight

Foot bracket (1)

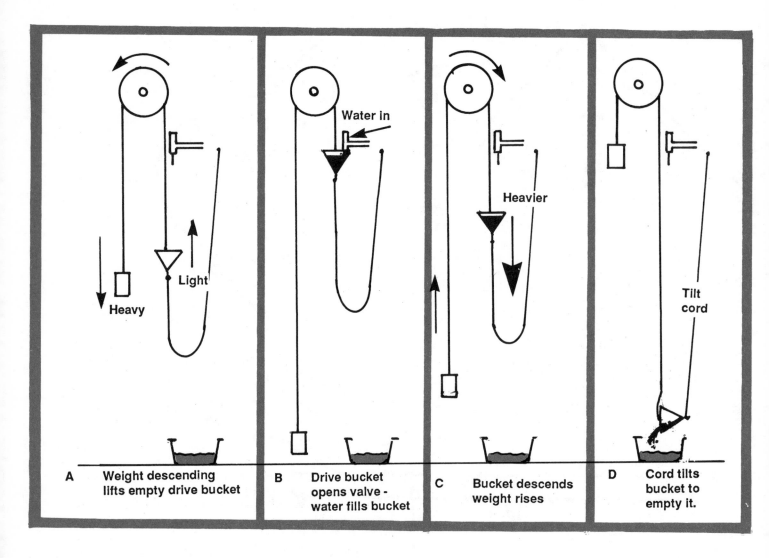

A	**Weight descending lifts empty drive bucket**
B	**Drive bucket opens valve - water fills bucket**
C	**Bucket descends weight rises**
D	**Cord tilts bucket to empty it.**

Step 3 Use corrugated cardboard for the well roof. After cutting this piece out with scissors, score it across the corrugations so that it can be bent. Cut a paper ridge strip and score this, also.

Step 4 Trace the shapes of the well front and the man onto greaseproof paper. Transfer the outlines onto the tile by pricking just outside the line with a cocktail stick. This will leave holes large enough to be seen. Join up the holes with a pen line and cut fractionally inside to avoid the marks and holes.

When you have cut out the parts with the knife tip, draw round the leg and arm to get a pair of each. Cut these out, put each pair together and pierce them for pivots.

With scissors, cut out four washers from the flat plastic of a margarine tub lid.

Step 5 Using pliers, reshape a giant paper clip to form a winding shaft. Lay it in the plan to check the shape.

From a cotton bud stem, cut four short bearings and one long one. Keep these parts together.

Cut the bottom foot from the yoghurt pot. Do not cut into the grooved part while removing the pot. Trim it down with scissors, until it matches the drawing.

With luck, there will be a moulding mark in the centre, otherwise use the centre finder or a folded paper strip to mark it there. Pierce the hole with a pin, then enlarge it to take the wire. Cut a scrap of balsa wood, pierce it and slide it on to support the centre. Epoxy the wood to the plastic. This forms the pulley. Test spin it with the wire in place. It should not wobble. Remove any tilt, to correct.

Step 6 Make the drive bucket like a cone, so that the water will run out when tilted. Cut a piece of plastic from the lid of one of the margarine tubs. Roll it to a point and epoxy the edges together. Hold with masking tape until set. Pierce and fit a paper clip handle. Epoxy a pin at the bottom and squeeze a scrap of modelling clay into the point.

Step 7 Make a water release valve, as follows. Pierce a hole in a bendy straw ½" (6mm) from the end. It has to fit a thin straw. Cut a piece of cotton bud stem, making sure the end is smooth and level.

Step 8 This is optional. The other bucket is non-funtional, so a small lump of modelling clay will do. The drawing explains.

Step 9 Place the valve through the hole in the base, and the tub behind it. Epoxy the back, screen, sides, foot bracket, well side and well front to the base. All are positioned as in Step 1.

Epoxy two short bearings into the back and well front. Slide the wire shaft in from the front, through the pulley, and through the rear bearing. With the handle up, bend the other end down a little and add a little clay to balance it. Do not put the roof on the well yet.

Put the model on a table, with the well overhanging the edge. Add clay to the back to steady it. Tie some of the thread to the working bucket, pass it over the pulley and down again. The bucket should be 4″ (100mm) from the floor. Tie a few knots in the other end near the base and put a small ball of modelling clay on it. Tie another piece of thread to the bottom of the bucket and fix it to the base so that the bucket is tilted.

Step 10 Push the bearing tubes into the body of the man and the foot bracket. They should grip, but if they are loose, epoxy them. While doing this, plug the ends with modelling clay so that they do not get clogged with epoxy. Clean them out afterwards.

Push a pin through the leg, a washer, the tube, another washer and the other leg. Epoxy the point and head to the respective legs. Do this at hips and feet. Both the ends should move freely.

Epoxy the foot bracket to the base. Epoxy the arms to the body at the angle shown. Pass the long bearing through the hands and epoxy it at exactly 90° to the surface. Lift the thread off the pulley.

Test spin the crank. Add clay to the rear end to balance the weight of the man. Replace the thread on the pulley and glue the roof on.

Testing
Put the other margarine tub on the floor under the bucket. Adjust the amount of clay on the thread so that it only just raises the bucket. Pull the bucket down and let go. It should be travelling fast at the top, striking the valve.

Fill the top tank and water should fill the bucket and send it down. At the bottom, the tilting thread should tighten and upset the water into the bottom tank.

The sequence is shown in 11. It may be necessary to put a plastic sheet guide plate under the base to make the bucket arrive squarely at the valve. Avoid tangles by attaching the tilt thread well away from the valve.

Bind it with a narrow strip of masking tape until it is a tight fit in the end of the straw. Drop a glass-headed pin down the bendy straw to stick out through the cotton bud stem.

Slide the thin straw into the hole in the bendy straw far enough to stop the pin sliding out. The pin will still lift up a little. Pierce one margarine tub at one end near the bottom corner, slide the other end of the thin straw in and epoxy it to the tub and bendy straw. When it is set, cut the bendy straw down level with the top of the tub.

Test the valve by putting water in the tub and gently pressing the pin point up. It may not completely shut off the water when you take the finger away, but there should only be one or two drips in, say, five seconds.

If it leaks more than this, pull the pin down, dragging the cotton bud stem out with it. See that the pin head sits on the tube evenly. Use another pin if the head is not a good fit. A very small smear of margarine on the tube end will also help. Push the valve tube and pin back again. Remember, the pin should not rise more than the distance on the drawing. If it does, push the valve tube in further. Empty the tub.

5 Aqua Pod

WATER-POWERED vehicles tend to be rather heavy, because the amount of water needed to drive them a respectable distance weighs so much.

This model, however, carries no water; it orbits around the power plant that drives it. A long spinning shaft transmits the power from a water wheel 'turbine', itself twisting around under a feed tank. Water then goes into a fixed waste tank.

Step 1 Trace the outline of the pod body side (B1) onto greaseproof paper, and don't forget the position of the bearing hole, which is the radius point for the top arc. Prick through just outside the line onto a tile and join up the marks with the pen.

Measure out and mark the ends B2 and B3 on the tile. Cut these three parts out with the modelling knife. Cut the card facings, and also a strip of paper to go the full length of the top.

Draw around B1 on another tile and cut it out. Place the sides together and pierce both at the same time with a cocktail stick. Epoxy the other facing outside.

Step 2 With scissors, cut the wheel facings from card. Epoxy one in from the edge of the third tile, so

Tools

● modelling knife ● straight edge ● ruler
● scissors ● pliers ● compass
● glasspaper ● fine felt pen

Materials

● 3 tiles ● postcard ● a yoghurt pot, cottage cheese tub, or other similar thin container with a bottom 2¼" (56mm) diameter ● 3 1-litre, round ice cream tubs ● a metal screw cap from a drinks bottle ● a piece of ¼" (6mm) diameter hardwood dowel 10½" (263mm) long ● a cap from a felt tip pen (to fit easily on the dowel) ● a 36" length of ⅛" (3mm) square section hard balsa wood strip (this can be cut from balsa sheet, provided it is long enough) ● a cocktail stick ● a cotton bud stem ● an empty ball-point pen ink tube ● a thin straw ● a bendy straw ● bicycle valve rubber ● 2 giant paper clips ● masking tape ● greaseproof paper ● thick paper ● sewing thread

Glue with fast epoxy

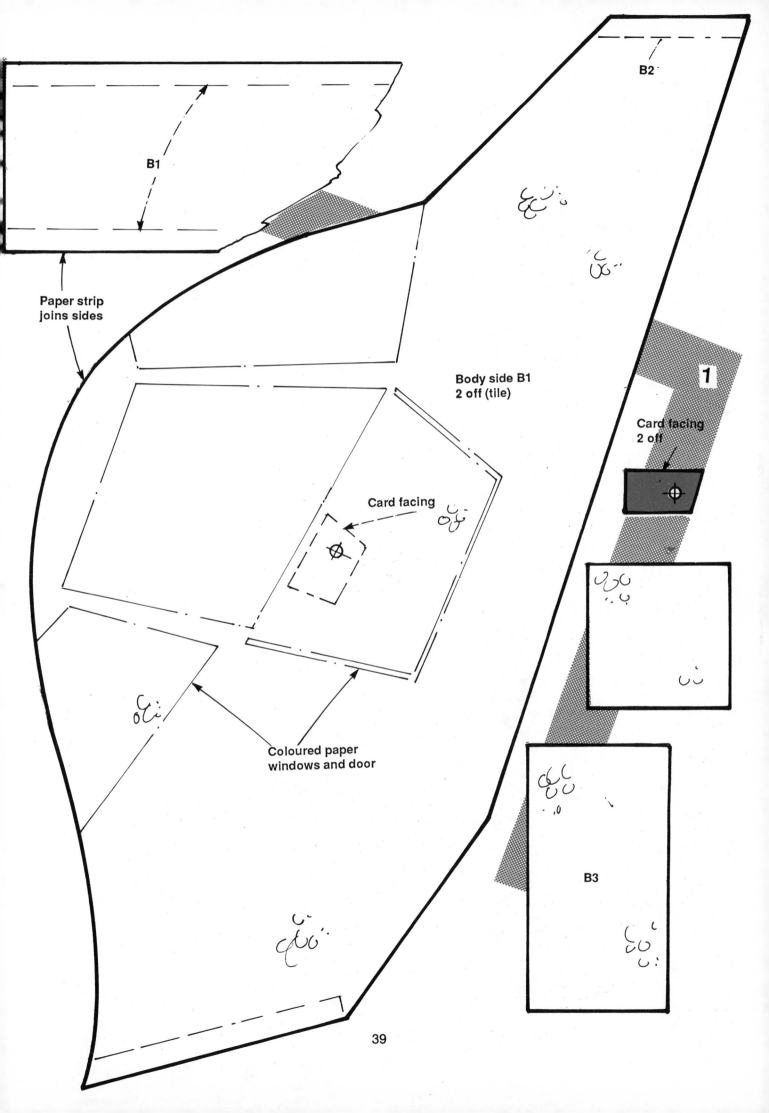

B2

B1

Paper strip
joins sides

Body side B1
2 off (tile)

Card facing

Coloured paper
windows and door

1

Card facing
2 off

B3

39

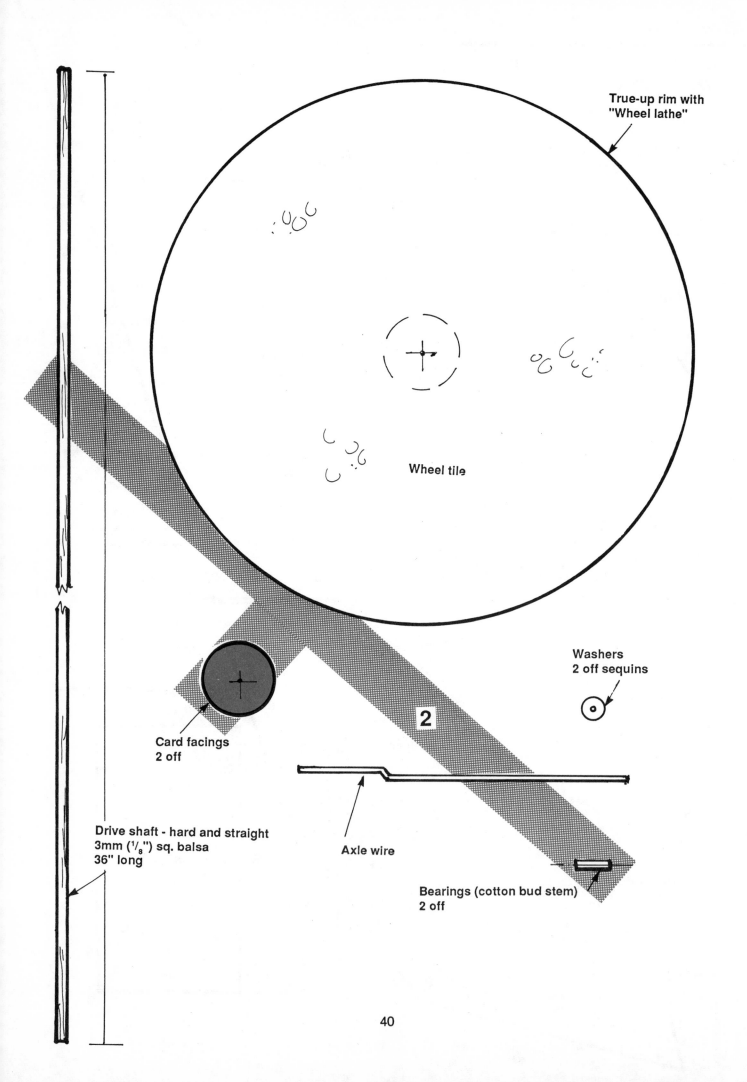

True-up rim with "Wheel lathe"

Wheel tile

Washers
2 off sequins

Card facings
2 off

2

Axle wire

Bearings (cotton bud stem)
2 off

Drive shaft - hard and straight
3mm ($^1/_8$") sq. balsa
36" long

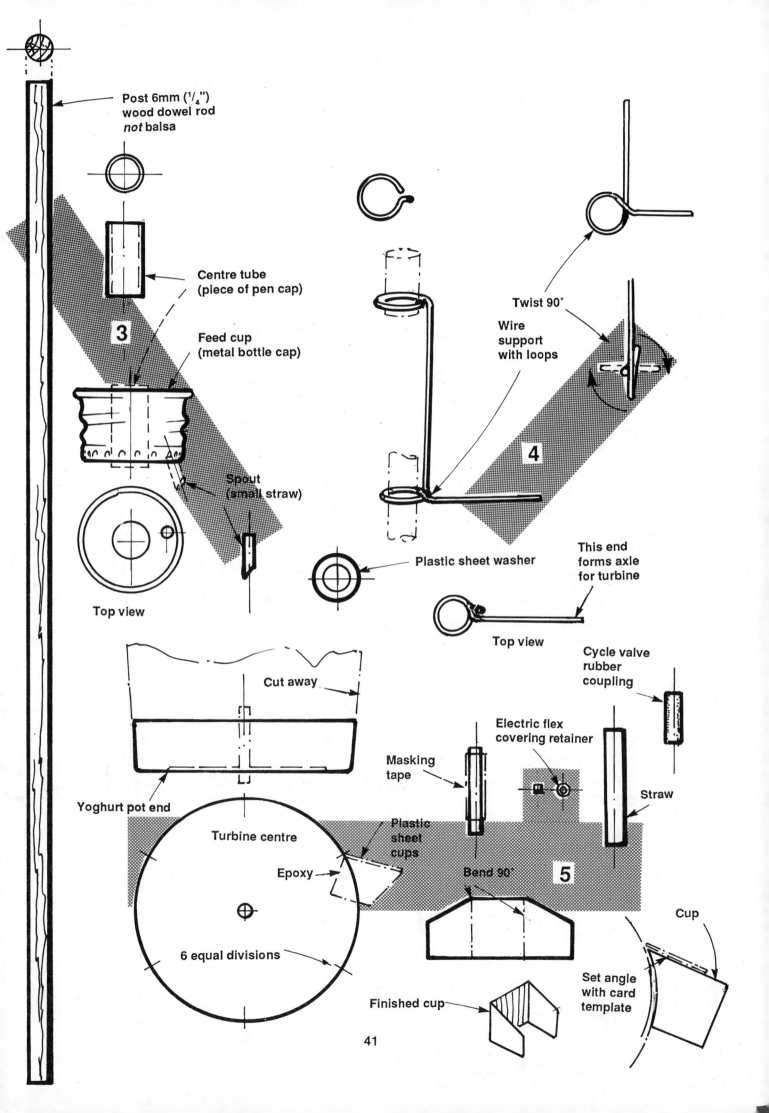

Post 6mm ($\frac{1}{4}$")
wood dowel rod
not balsa

Centre tube
(piece of pen cap)

3

Feed cup
(metal bottle cap)

Spout
(small straw)

Top view

Twist 90°

Wire
support
with loops

4

Plastic sheet washer

This end
forms axle
for turbine

Top view

Cut away

Yoghurt pot end

Turbine centre

Epoxy

Plastic
sheet
cups

6 equal divisions

Finished cup

Masking
tape

Electric flex
covering retainer

Bend 90°

5

Cycle valve
rubber
coupling

Straw

Cup

Set angle
with card
template

41

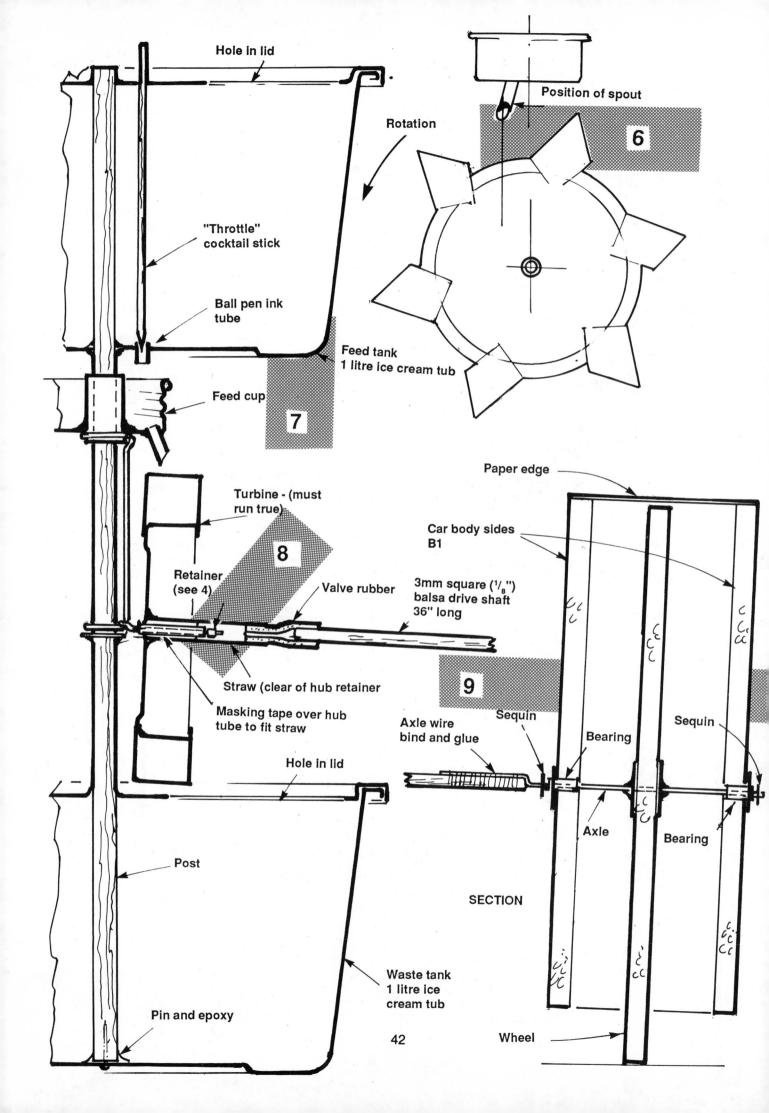

Hole in lid

Rotation

"Throttle" cocktail stick

Ball pen ink tube

Feed tank 1 litre ice cream tub

Feed cup

7

Position of spout

6

Turbine - (must run true)

8

Retainer (see 4)

Valve rubber

Straw (clear of hub retainer

Masking tape over hub tube to fit straw

Hole in lid

Post

Pin and epoxy

Waste tank 1 litre ice cream tub

Paper edge

Car body sides B1

3mm square (¹⁄₈") balsa drive shaft 36" long

9

Axle wire bind and glue

Sequin

Bearing

Sequin

Axle

Bearing

SECTION

Wheel

42

that a 5 ¾" (150mm) diameter circle can be drawn from its centre point. When set, mark out with the compass and cut with the modelling knife, in short slices, just outside the line. Pierce the centre with a pin, then with a paper clip. Epoxy the opposite facing in place. Check that it spins true on the wire. Remove the wire and allow to set.

Insert a wire or pin and finish the edge of the wheel on the 'wheel lathe', or by hand, using glasspaper at the edge of the table. Form the axle from a paper clip and make sure it is quite straight except where shown. Lay it on the plan to check. Cut two bearings from a cotton bud stem and epoxy them in the pod sides.

The drive shaft is simply a length of strip balsa which is straight and hard.

Step 3 Smooth the ¼" (6mm) hardwood dowel to form a supporting post for the feed tank. Find a cap or casing from a felt pen that will slide loosely over the dowel. Slice off 1" (25mm) to form a tube. Make a central hole in the thin metal lid of a soft drinks bottle. Start with a pin, enlarge with a cocktail stick, then rotate on the tip of a ball pen or other tapered item to make the hole fit the tube just made.

Make another small hole near the edge to take a short piece of thin straw. Epoxy both the tube and straw into the cap as shown. Note that the centre tube projects slightly below.

Step 4 Form a paper clip into a rotating support by shaping an upper loop to fit on the tube just mentioned, and a lower loop to slide on the dowel. The end extends to make an axle for the turbine. Epoxy the top loop to the bottom of the tube.

Step 5 Make the turbine, as follows. Cut the bottom from the yoghurt pot leaving ½" (12mm) of side on it. This forms the centre. Pierce it exactly in the middle for a cotton bud stem. Bind this with masking tape to wedge in a slim straw. Cut a small piece of insulation from electrical wire to fit the axle. Use as a retainer.

From the rest of the yoghurt pot, flattened out, cut six pieces to form water cups. Bend as shown and, using a card angle template, epoxy them equally spaced around the yoghurt pot bottom.

Step 6 Epoxy a thin straw into the turbine centre, at 90° to the bottom. Slide the tape-wrapped cotton bud bearing into the straw. Slide the turbine onto the shaft and test spin it. Adjust if necessary to take any wobble out, and trim any cups that stick out too far. Adjust the spout to the position shown.

Step 7 Take two of the 1-litre ice cream tubs. Pin through the centre of the bottom of one into the end of the hardwood dowel. Epoxy around it on the inside. This forms the post around which the turbine axle rotates. Pierce the lids of both tubs in the centre to take the post, and make a larger hole to one side in each.

Pierce the bottom of the other tub to fit the dowel and make a small hole near this for a piece of ink tube from a ball pen. This will feed water into the feed cup, which rotates below.

Pass the bottom lid down the post and put it on the tub. Cut a washer from more yoghurt pot plastic and thread it down to the position shown. Epoxy it from below.

Slide the turbine support wire down onto the washer. Check that it slides round the post freely, and does not rise and fall. (If the washer was not level it would do this.)

Slide the other ice cream tub (feed rank) onto the post and test that the fed cup does not rub on it. Epoxy the post to the tub inside only. Epoxy the ink tube in place. Put the lid on, with the post through it. Make a small hole near the post for a cocktail stick to reach into the ink tube. This is the throttle.

Step 8 Remove the cotton bud bearing from the straw in the turbine. Slide the bearing onto its axle. Slide the retainer on to stop the bearing sliding off. Put a tiny dab of epoxy on the end to hold it there (it should still spin). Slide the turbine back onto the bearing. It should grip without epoxy.

Step 9 Bind and epoxy the wheel axle to the drive shaft. Test spin, and correct any wobble.

Slide on a sequin, followed by one pod side, the wheel (which you epoxy to the centre), then the outer side, and another sequin.

Test spin the wheel via the shaft to check that the wheel is true on the axle. Adjust before the epoxy goes hard.

Epoxy the sides to B2 and B3 and hold with pins to keep square. Cover the top with a strip of paper to complete the pod vehicle. Do not paint it, and keep it light. If the pod body tips up, add a very small piece of modelling clay to the bottom edge to balance it on the axle.

Push a piece of valve rubber onto the other end of the shaft, and plug it into the turbine straw. Prop the pod wheel off the floor and spin the turbine to see that it runs smoothly. The turbine should be in line with the shaft, not at an angle to it.

HOW IT WORKS
Carry the model by holding the power unit in one hand and the pod in the other. Remember that the shaft is light and easily broken (but easily replaced). Put the model on a smooth level surface and fill the supply tank. Lift the throttle a little.

The turbine should start to turn. If it's sluggish, gently push start the pod, which should then drive round and round, making the turbine turn around the post as it goes. The water runs from the turbine into the lid of the waste tank, then through the hole into it.

When the water is used up, unplug the valve rubber from the straw, and tip the water into the spare ice cream tub. Stand it the right way up, re-couple the valve rubber and refill the top tank from the spare tub.

6 Twin Tilter

THIS IS a novel type of engine which uses water as the power supply. The prototype runs for about ten minutes on a 16oz (500g) container of water. It turns around 60 revs per minute, so that you can see the action clearly, and gives twin power pulses – that's 120 power strokes per minute.

There is a throttle control, so that it can be made to run faster or produce more power to drive other displays for a slightly shorter time. Of course, a larger supply container can be added to make it run for longer.

Step 1 Transfer the shapes of the frame base and top to the tile, ¼″ (6mm) from the edge to avoid the chamfered tile edge. Two of each part will be needed, so mark them out close to each other. Place the frames together and pierce the holes through both with a cocktail stick, so that they line up when assembled later.

Step 2 Make two bearings from cotton bud stems. Note that one is shorter than the other – this is the

Tools

● modelling knife ● straight edge ● pliers
● ruler ● scissors ● compass ● fine tip felt pen

Materials

● 2 tiles ● a 1lb (500g) size rectangular margarine tub ● two 8oz (250g) plastic tubs
● strips of thin plastic cut from a large size container (washing-up liquid bottles will do)
● small pieces of postcard or cereal packet
● 4 giant size paper clips ● 2 cotton bud stems
● 6 sequins ● cocktail stick ● short piece of small diameter drinking straw or piece of used ink tube from a ballpoint pen ● small piece of modelling clay ● pins ● masking tape

● Glue with fast epoxy

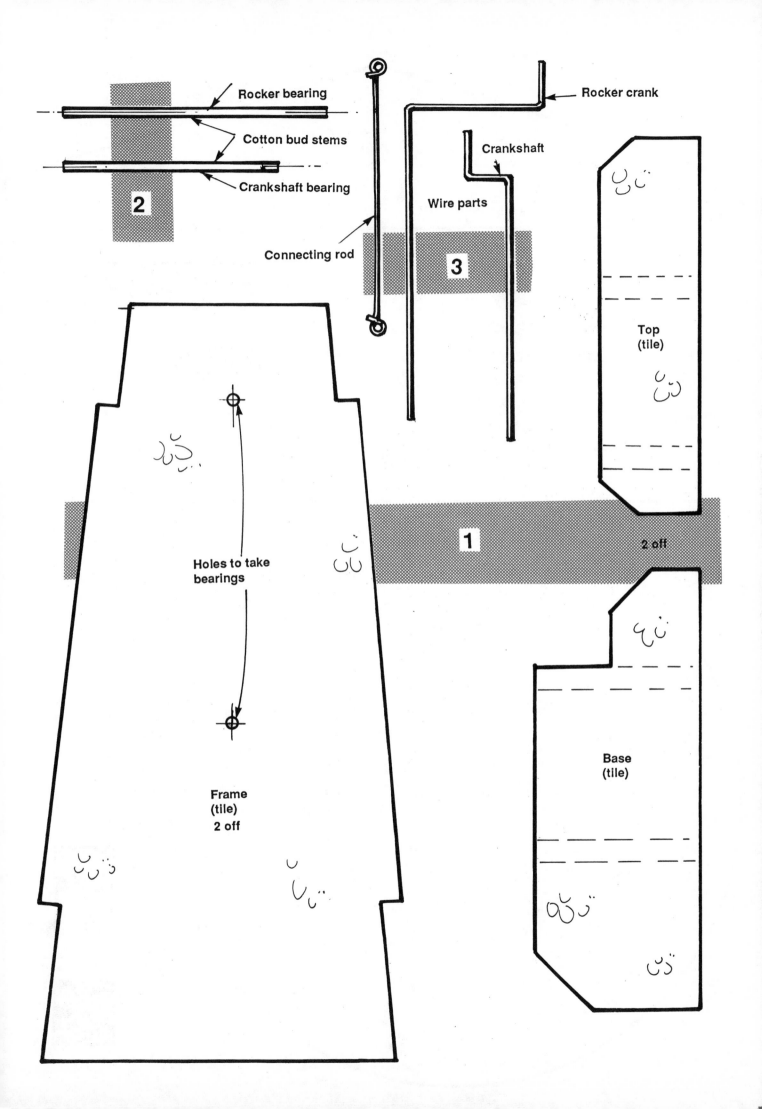

Rocker bearing

Cotton bud stems

Crankshaft bearing

2

Rocker crank

Crankshaft

Wire parts

3

Connecting rod

Top (tile)

1

2 off

Holes to take bearings

Frame (tile) 2 off

Base (tile)

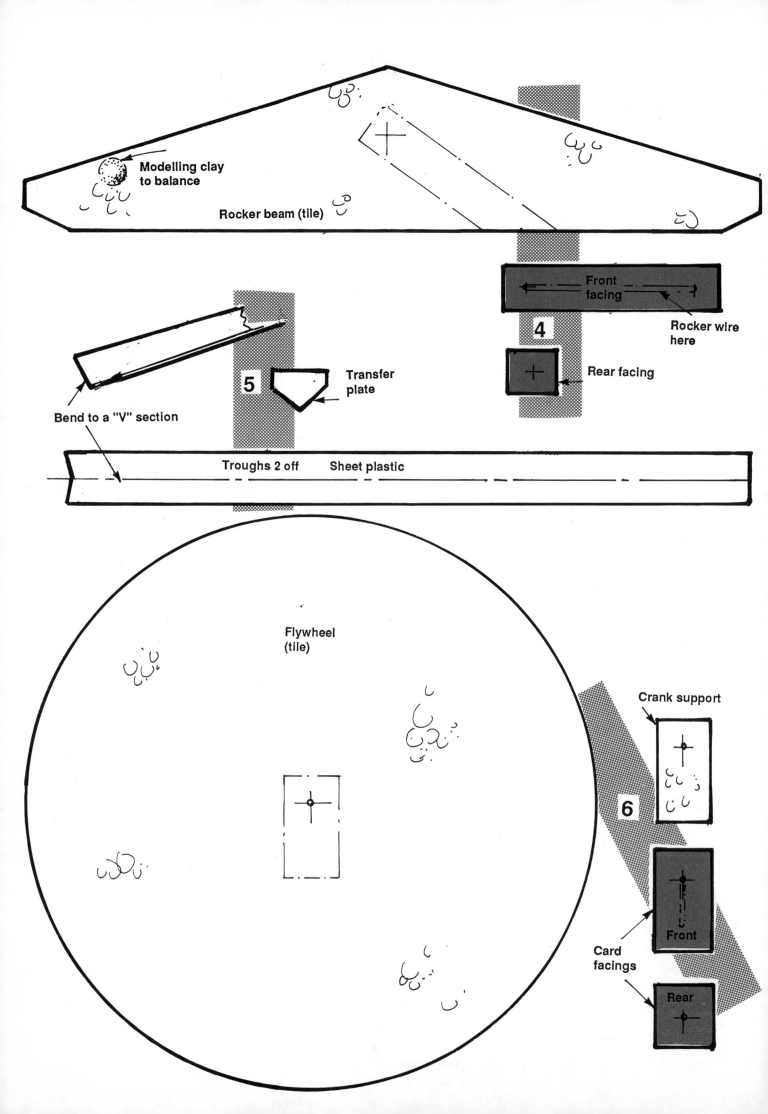

Modelling clay to balance

Rocker beam (tile)

Bend to a "V" section

Transfer plate

Troughs 2 off   Sheet plastic

Front facing

Rocker wire here

Rear facing

Flywheel (tile)

Crank support

Front

Card facings

Rear

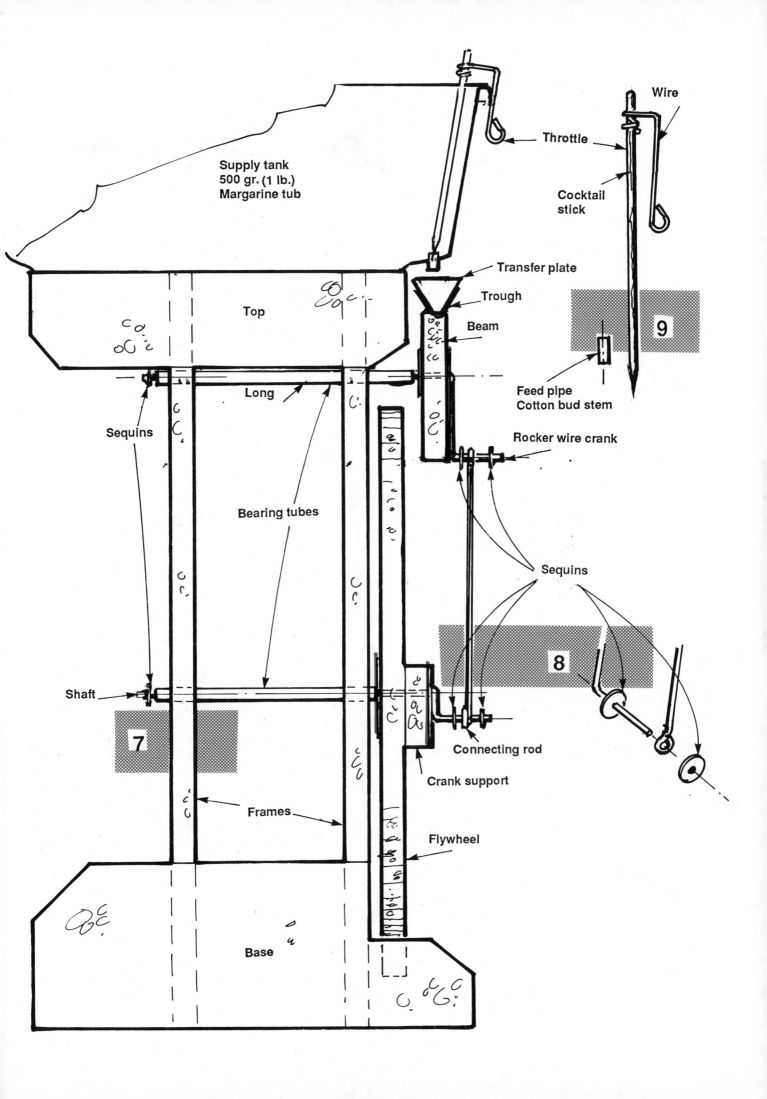

Supply tank
500 gr. (1 lb.)
Margarine tub

Top

Sequins

Long

Bearing tubes

Shaft

7

Frames

Base

Transfer plate

Trough

Beam

Feed pipe
Cotton bud stem

Rocker wire crank

Sequins

8

Connecting rod

Crank support

Flywheel

Wire

Throttle

Cocktail
stick

9

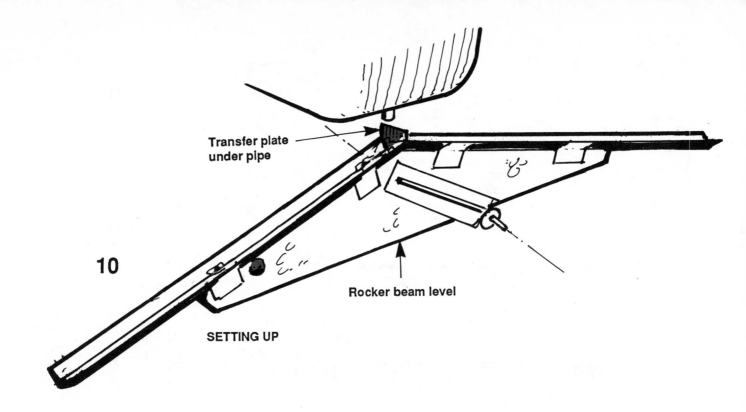

Transfer plate under pipe

Rocker beam level

10

SETTING UP

11 SEQUENCE OF OPERATION

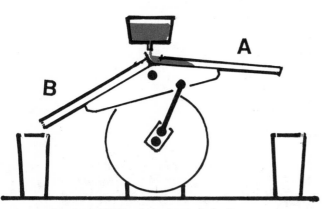

A   Start-up. Water into trough A. Trough B empty

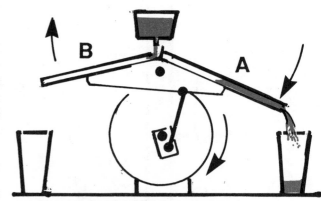

B   POWER STROKE 1. A is heavier than B so beam tilts - water empties, feed transfers to B

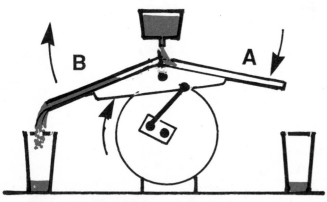

D   POWER STROKE 2. B is heavier than A - feed transfers, as beam tilts B empties

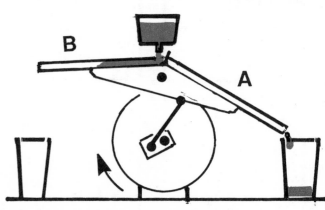

C   Trough B filling, trough A emptying

lower one to carry a crankshaft. There should be no cotton left on the ends.

Step 3 Straighten out three giant paper clips, checking that there are no kinks left, by rolling them on a flat surface.

Using the small nose of your pliers, form the bends as shown on the plan. Lay the parts over the drawing and adjust the bends and lengths until they match exactly. Paper clips are cheap, so do not attempt to rebend really mismatched wire parts.

Check that the crank wires turn freely in the tubes and that the loops at the ends of the connecting rod are easy, but not sloppy, fit on the crank wires. Put these parts aside.

Step 4 Transfer the shape of the rocker beam to the second tile and, having cut it out, pierce it with a cocktail stick, so that the rocker wire (Step 3) can be fitted.

Cut the card facing pieces and carefully pierce them so that the wire just passes through. Mark the position of the larger piece on the front of the rocker beam.

Step 5 Water has to run along two troughs which go on the top of the rocker beam, so these have to be light and waterproof. The ideal material is the thinnest part of a margarine tub cut diagonally across the lid, or around the sides. Measure the length with a strip of paper so, if the tub is too small, it need not be wasted. You can also use thin washing-up liquid bottles – cut a couple of strips from top to bottom.

Shape one end of each strip and, using a ruler, fold the strips lengthwise to form a 'V'-sectioned trough. A small piece of margarine tub plastic forms a 'transfer plate' which divides the troughs from each other. Put these parts aside with those from Step 4.

Step 6 The engine is so efficient that, unless it has to do fairly hard work, the flywheel need not be heavy. A piece of tile is fine for the job. Should it need to be weightier, then push small nails into its rim.

Stick a scrap of masking tape on the tile where the centre of the wheel is to be. Place a fine tip felt pen in a school compass and, measuring from the drawing, mark out the flywheel rim. The masking tape should prevent the centre point from tearing the tile.

Cut the crank support from the tile waste, and a matching facing piece from a postcard or cereal packet. A smaller facing from the same card goes behind the flywheel.

Pierce all these parts to fit the crank wire (Step 3).

Step 7 This is the main assembly, so work in the following sequence.

Collect the frame parts (Step 1). Mix up a small quantity of fast epoxy. Glue the base and top pieces to the frames. Pins or masking tape will hold them true while the adhesive sets.

Immediately slide the long upper bearing and shorter crankshaft bearing (Step 2) through the frame. Without getting epoxy inside the ends, put a touch of it where it passes through, slide the tubes in and out slightly and rotate them to spread the epoxy into the joints. Refer to the side view for the amount of tube projecting. This should take less than five minutes, or the epoxy will have hardened.

Mix a very small amount of epoxy and fix the crank support to the flywheel and the card facings in place. (Step 6). Slide the crankshaft wire through from the front and epoxy it to the card facings. Wipe surplus epoxy from the shaft, so that it will run smoothly on the tubes. Do not place it in the tube while the epoxy is wet. Prop the assembly upright, and make sure that the shaft is at 90° to the face of the flywheel.

Using epoxy and pins, fix the troughs (Step 5) to the beam (Step 4). Insert the transfer plate between the ends of the troughs and seal the joint and around the pin heads with more epoxy. Masking tape will keep the trough upright on the edge of the beam while the epoxy sets.

Epoxy the card facings to the beam and slide the rocker crank (Step 3) through. Epoxy this to the card facings, and wipe excess epoxy off the wire. Slide the crankshaft assembly into its tube and wedge a sequin onto the rear end to secure it. Leave enough end-to-end movement to allow it to spin freely.

Step 8 Insert the rocker crank wire into the upper tube and retain it with a second sequin. Check that it rocks freely. Place a sequin on eack crank close to the bends.

Fit the connecting rod (Step 3). Push two more sequins onto the cranks to prevent the rod slipping off. Spin the flywheel to check that the engine runs smoothly. Add a small blob of modelling clay to the beam to balance the cranks. It should now stop wherever it is placed, as it has no preference.

Step 9 Form a throttle control valve by winding a giant paper clip around the top end of a cocktail stick. Adjust it to grip the edge of the supply tank, which is a margarine tub. It should reach the bottom. If it does not, join a second cocktail stick on by wedging the ends onto a short piece of cotton bud stem. The point of the stick fits loosely into the feed pipe, which is a short piece of very small diameter drinking straw (1/8" or 3mm). Alternatively, a piece of ink tube from a used ballpoint pen will do.

Epoxy the tube into a hole pierced close to the edge of the margarine tub bottom. Do not, at this stage, fix the tub to the rest of the model.

HOW IT WORKS

Place the model on a level surface, preferably outdoors or where slight water splashes are not important. Position the two 8oz (250g) containers under the ends of the troughs. Place the supply tank – the 16oz, 500g tub – on top of the engine. Check that, when the beam is level (equal distance between each end and the ground), the feed pipe is exactly over the transfer plate. It should look like Fig. 10.

Turn the flywheel to rock the beam to one side. Fill the tank with water and adjust the throttle without moving the tank. Water should start to run into one trough, which should now tilt down causing the water to feed the other trough. The sequence of operation is shown in Fig. 11.

If the action is uneven, slide the tank towards the slower side. Once set up, it may be fixed with epoxy to the frame.

To avoid unnecessary water splashes, close the throttle, or wait until the water is used up, before emptying the water back from the two smaller tubs into the supply tank.

7 Syphon Clock

THERE are many types of water clock, both simple and complex. One problem with the simple type is that the clock may slow as the water is used up. This example gets around this by using a floating syphon to give an even flow of water. As the water level goes down the pipe goes with it, so the pressure remains exactly the same.

A sliding paper scale is used as a 'dial', to ensure that the timing is correctly measured and set.

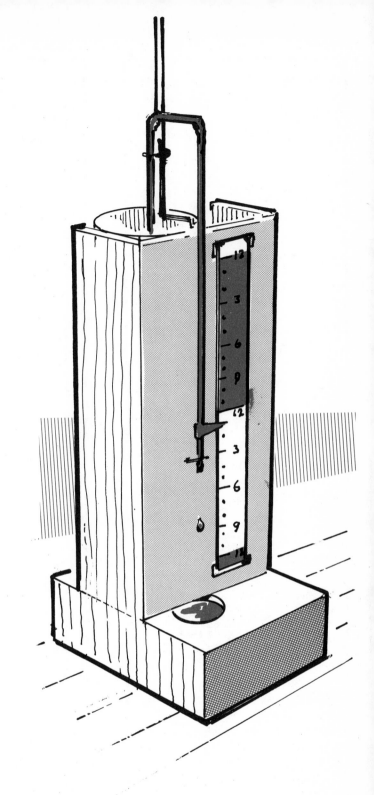

Tools

- modelling knife ● straight edge ● pliers
- scissors ● ruler ● fine felt tip pen
- ballpoint pen

Materials

- 2 tiles ● a bendy straw ● 1 normal size straw ● 3 slim straws ● a cotton bud stem ● 4 giant paper clips ● 2 small paper clips ● paper ● thin rubber band ● a table tennis ball ● a 500g size washing-up liquid bottle ● a 500g size rectangular margarine tub ● cycle valve rubber tube

Glue with epoxy

Step 1 Mark out and cut from tile the face, front and two upper sides. The slots top and bottom in the face must be exactly parallel to the edge.

Step 2 Using more tile, cut two lower sides, the top and spacer.

Step 3 Straighten a giant clip, shorten it and slide it into a slim straw just past centre, It can be prodded along with another piece of wire. Bend the straw at this point, to the shape shown.

The wire will bend with it and hold the shape. Water will still pass through the straw, even at the crumpled part.

Gently bell out one end of the other slim straws, using the point of a pencil or ball pen, so that they slip over the bent straw for about ³/₁₆" (4.5mm). Epoxy these joints, lay over the drawing and adjust the angle and length of each side. This forms the syphon.

Cut a short end piece of straw, and bell it to fit dry (not glued) onto one end of the syphon. Make a clamp from a giant clip, to squash the end piece. It has to be doubled over very tightly to make the water drip slowly enough.

Finally, make a figure-eight shaped guide to grip the syphon near the top. Use a small clip.

Step 4 Bend a 3-point guide from a giant clip. This keeps the syphon centrally in the main tank. Drill a table tennis ball right through for the syphon straw. Epoxy the straw into the ball with a small projection below. Epoxy the guide to this point, as shown.

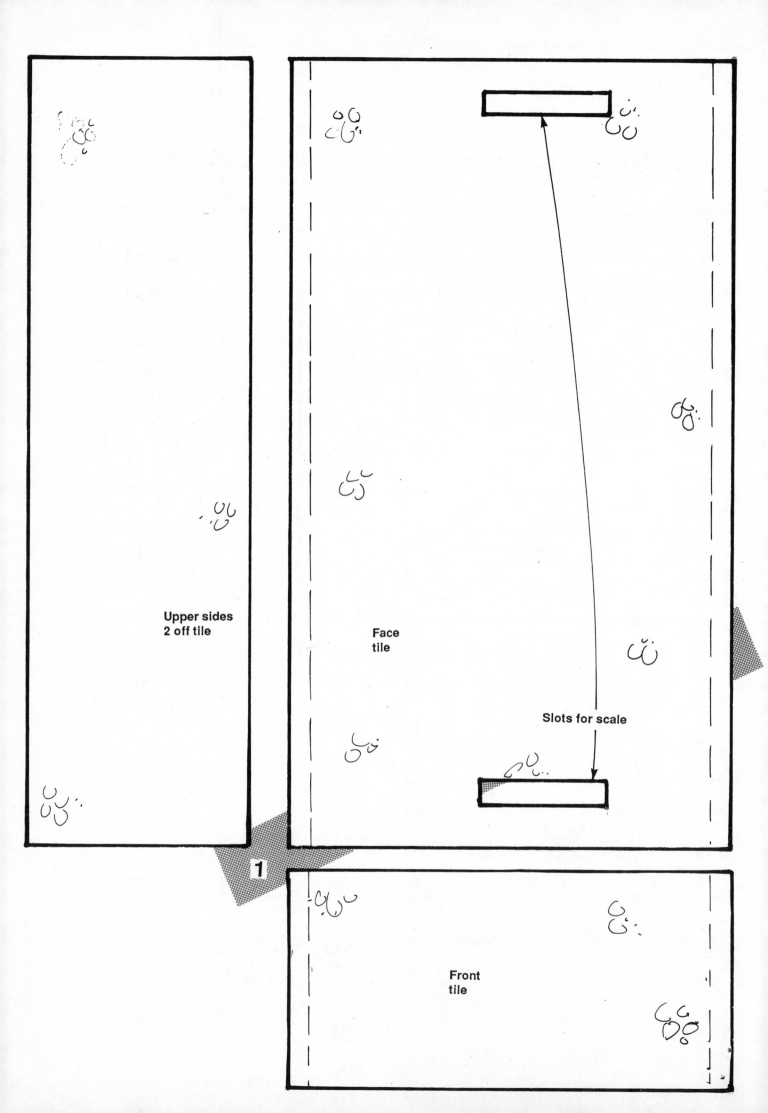

Upper sides
2 off tile

Face
tile

Slots for scale

1

Front
tile

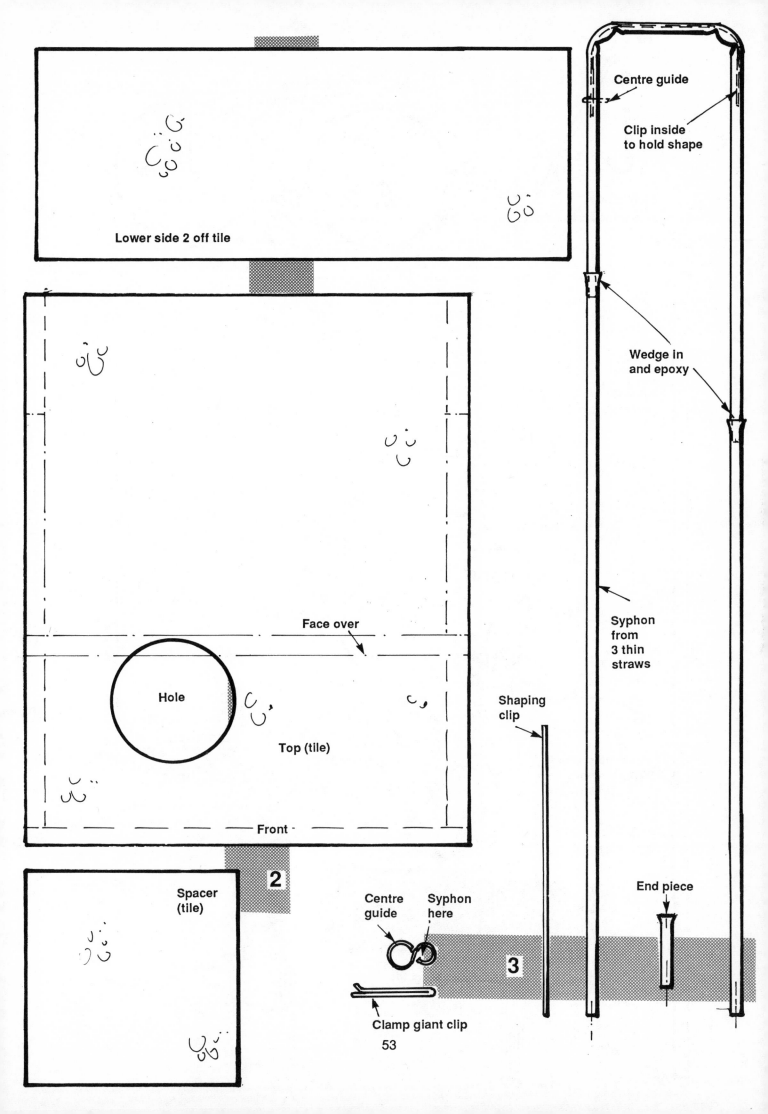

Lower side 2 off tile

Face over

Hole

Top (tile)

Front

2

Spacer (tile)

3

Centre guide

Syphon here

Shaping clip

Clamp giant clip

Centre guide

Clip inside to hold shape

Wedge in and epoxy

Syphon from 3 thin straws

End piece

53

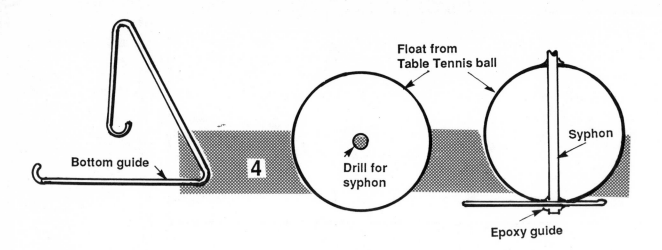

Bottom guide

4

Float from
Table Tennis ball

Drill for
syphon

Syphon

Epoxy guide

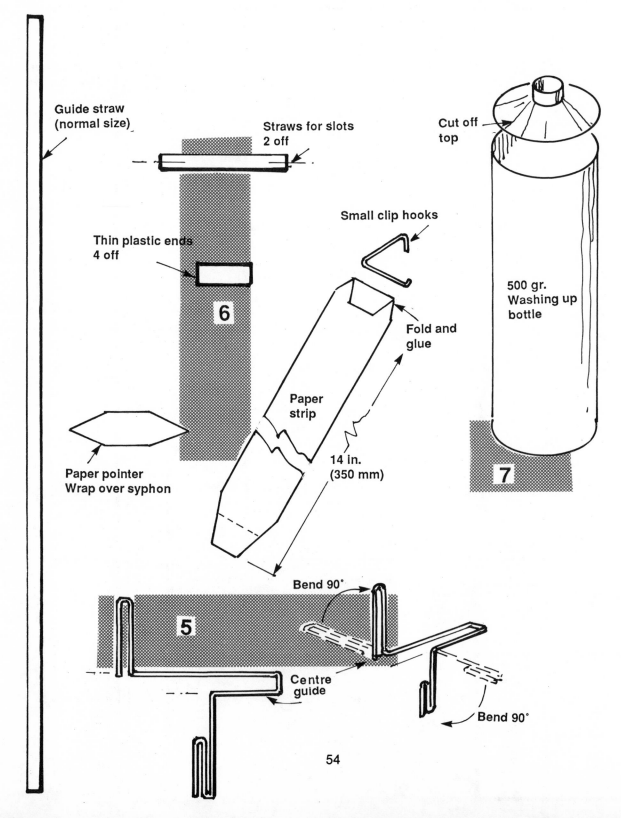

Guide straw
(normal size)

Straws for slots
2 off

Thin plastic ends
4 off

6

Small clip hooks

Fold and
glue

Cut off
top

500 gr.
Washing up
bottle

7

Paper
strip

14 in.
(350 mm)

Paper pointer
Wrap over syphon

Bend 90°

5

Centre
guide

Bend 90°

54

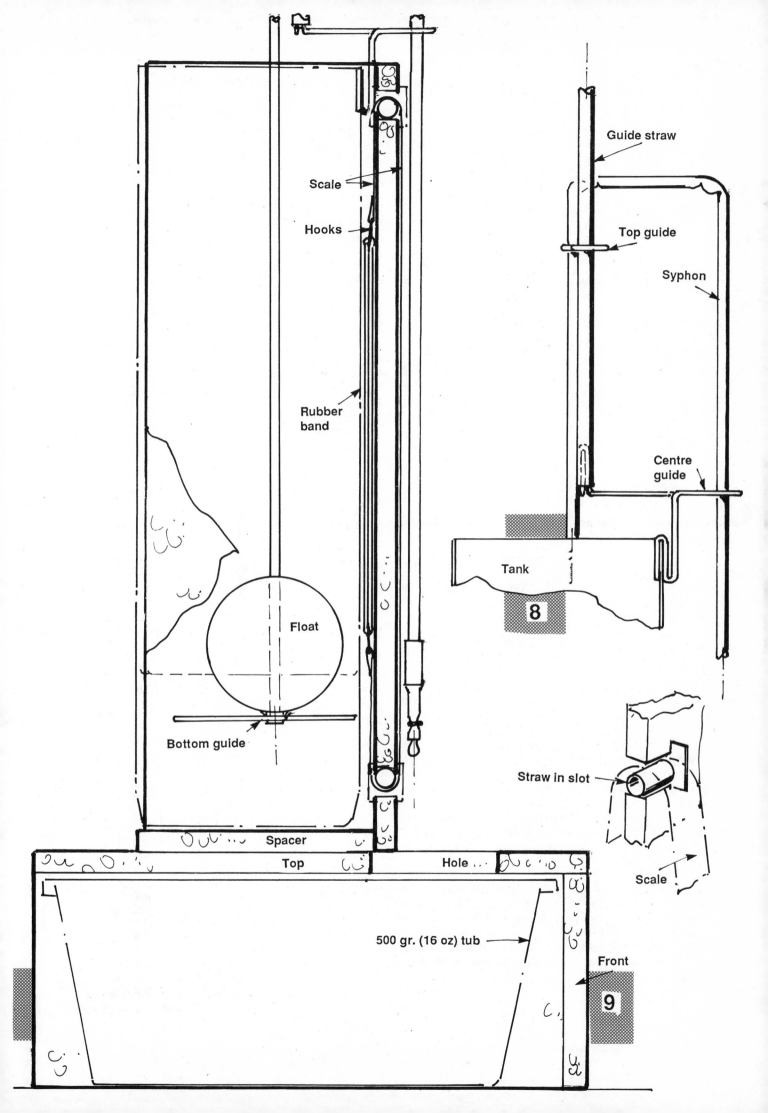

Scale

Hooks

Rubber
band

Float

Bottom guide

Spacer

Top

Hole . . .

500 gr. (16 oz) tub

Guide straw

Top guide

Syphon

Centre
guide

Tank

8

Straw in slot

Scale

Front

9

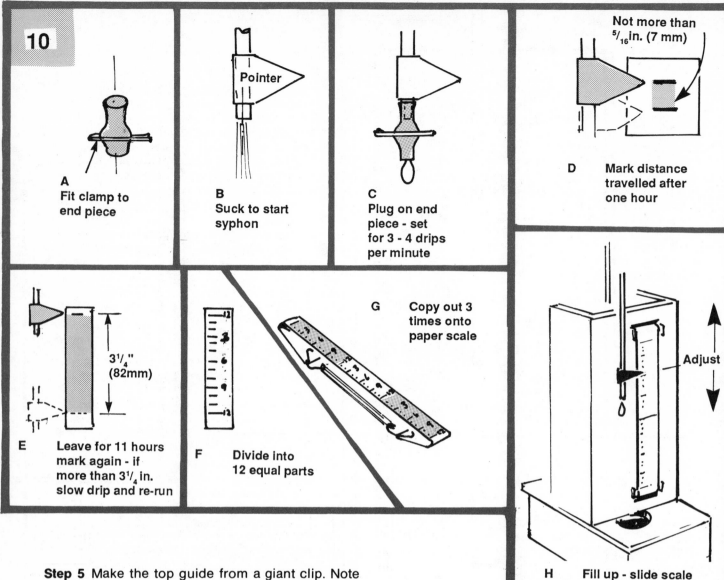

10

A Fit clamp to end piece

B Suck to start syphon

C Plug on end piece - set for 3 - 4 drips per minute

D Mark distance travelled after one hour

Not more than 5/16 in. (7 mm)

E Leave for 11 hours mark again - if more than 3¹/₄ in. slow drip and re-run

3¹/₄" (82mm)

F Divide into 12 equal parts

G Copy out 3 times onto paper scale

H Fill up - slide scale to match time against pointer

Adjust

Step 5 Make the top guide from a giant clip. Note that its end parts are twisted so that they point up and down, leaving the centre loop flat, to go around the syphon. Epoxy a standard straw into the upper loop to support the top of the syphon over the main tank.

Step 6 Cut the bendy straw to make two smooth facings to fit the slits in the face. Scraps of thin plastic (margarine tub lid) go at either end to tidy it up.
The time dial is a long paper strip, which will be threaded through the slots, over the straws and connected by a rubber band behind the face. Small wire hooks carry the band.

Step 7 Cut the conical top from a 500g washing-up bottle, leaving a tubular main tank. The top, minus its nozzle, will make a convenient funnel for filling the tank later.

Step 8 Wedge the top guide onto the edge of the main tank, with the loop outside. Drop the syphon and ball into the tank, threading the outer end of the syphon through the top guide and the centre guide loop over the guide straw. It should slide up and down freely. Adjust the guides to ensure that this happens.

11 If left stopped put end in pot of water to halt system

56

Step 9 Using epoxy, assemble the upper and lower sides, the face and front and spacer to the top piece, using the position reference in that piece (Step 1). Slide the tank down with the outer end of the syphon hooked over the face. Do not glue it down yet.

Temporarily thread the paper scale through the slots and hook on the rubber band behind. Epoxy the pointer around the syphon ⅛" (3mm) from the end, and facing the scale.

Step 10 Now for setting up the clock and marking out the scale.

Almost fill the main tank. Place the waste tank (margarine tub) under the base. Continue as follows:

A) Slide the clamp onto the end piece of the syphon, but do not plug the end piece on yet.
B) Hold the valve rubber onto the end of the syphon and suck with the mouth until water flows out.
C) Plug the end piece on so that the clamp does not rub on the scale or dig into the face. Slide the clamp across until the rate of dripping is down to three or four drips per minute.
D) Mark the position of the pointer with a pencil. Leave the clock going for one hour, then mark the new position of the pointer. Expect it to be just over ¼" (6 or 7mm) below. If it is more, adjust the clamp to slow the drip rate, as the clock would run out of water in less than 24 hours.
E) If all is well, leave the clock going for 11 more hours. If the first drip rate check had to be repeated, to get it right, time from that last pair of marks. Either way, expect to measure about 3¼" (82mm) for the 12-hour total.
F) If this is so, pencil in 12 equal spaces.
G) Take the scale out of the slot and, using a ballpoint pen (so that the ink does not run), make a neat job of marking out three sets of equal spaces, to give a total of 36 hours in 12-hour blocks. Colour the centre one and mark it 'A.M.' Replace it in the face. Epoxy the tank to the side and top piece.
Take out the waste tank and empty the water into the main tank. Reposition the waste tank and top up the main tank until full.
H) Now note the actual time, and slide the marked scale up or down to match that number to the pointer. (That is why there are three sets of 12 hours on the scale.

Use this last adjustment to correct the time whenever necessary and when the clock is 'wound up', by tipping the water form the waste tank back into the main tank.

'Wind' at about the same time each day. The clock should run for a little over 24 hours, between 'windings' but do not allow it to run dry, or it will need restarting, as Step 10(B), although the clamp should not be altered while unplugging the end from the syphon and refitting it.

If the clock has to be stopped for a day or so, put a small pot of water under the dripping end, so that, on entering it, the syphon will stop but not empty itself.

Always keep the tank topped up after 'winding' to compensate for evaporation.

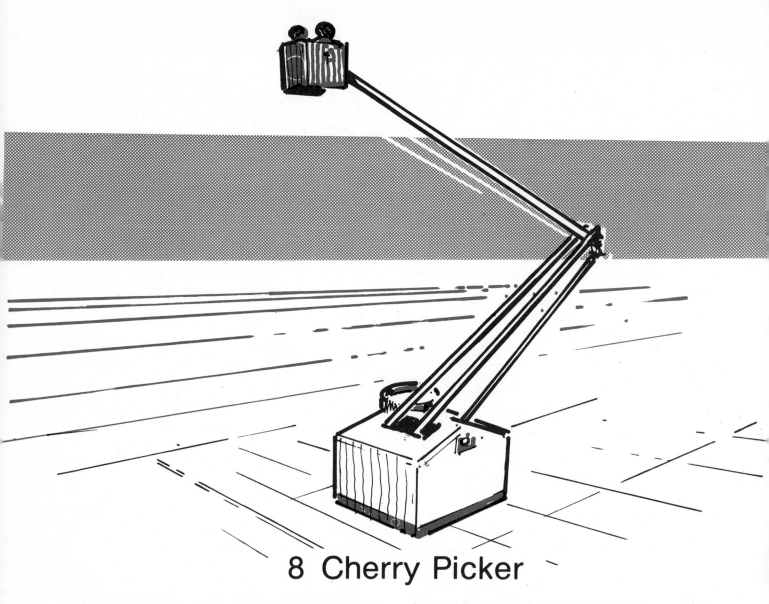

8 Cherry Picker

THIS MODEL uses hydraulic pressure to raise the long jointed arm. It is simply done and avoids the use of valves and proper hydraulic rams. It is self-contained, so does not need the water replacing. Set a small container of water on a raised bracket, and the arm slowly and steadily extends upwards. Put the container on the ground to make the arm fold down again.

Step 1 Cut four facings from card to reinforce around bearings and pivot pin. Mark out and cut from tile the base, front, back, sides (2), bracket and pivot plate. These will form the main casing.

Step 2 Cut from tile the shelf, its two supports, the pressure pad (which drives the arms) and the tank support.

Step 3 Cut the tile top pieces which cover the mechanism.

Step 4 The arms are made from straws. Pierce two where shown for a giant paper clip wire, which will form an axle, and at the other end with a pin.
  The top arm is single, pierced with a pin each end, and a third straw forms a link to control the top arm. This, too, has pin holes and is shorter.

Step 5 Bend a yoke to join the bottom ends of the main arm straws. It has an extension to form a drive horn. This has to be angled down as shown.
  Form a ball of modelling clay to the size shown, which should be about the right weight to partly balance the finished set of arms. Mould this into the yoke.

Tools

- modelling knife ● straight edge ● pliers
- scissors ● fine tip felt pen ● ruler

Materials

- 1 tile ● card ● 3 giant paper clips
- 1 small paper clip ● 4 standard straws
- hair spray lid ● cycle valve rubber
- standard round toy balloon ● modelling clay
- cotton bud stem ● sewing thread
- 4 sequins ● masking tape ●

Glue with epoxy

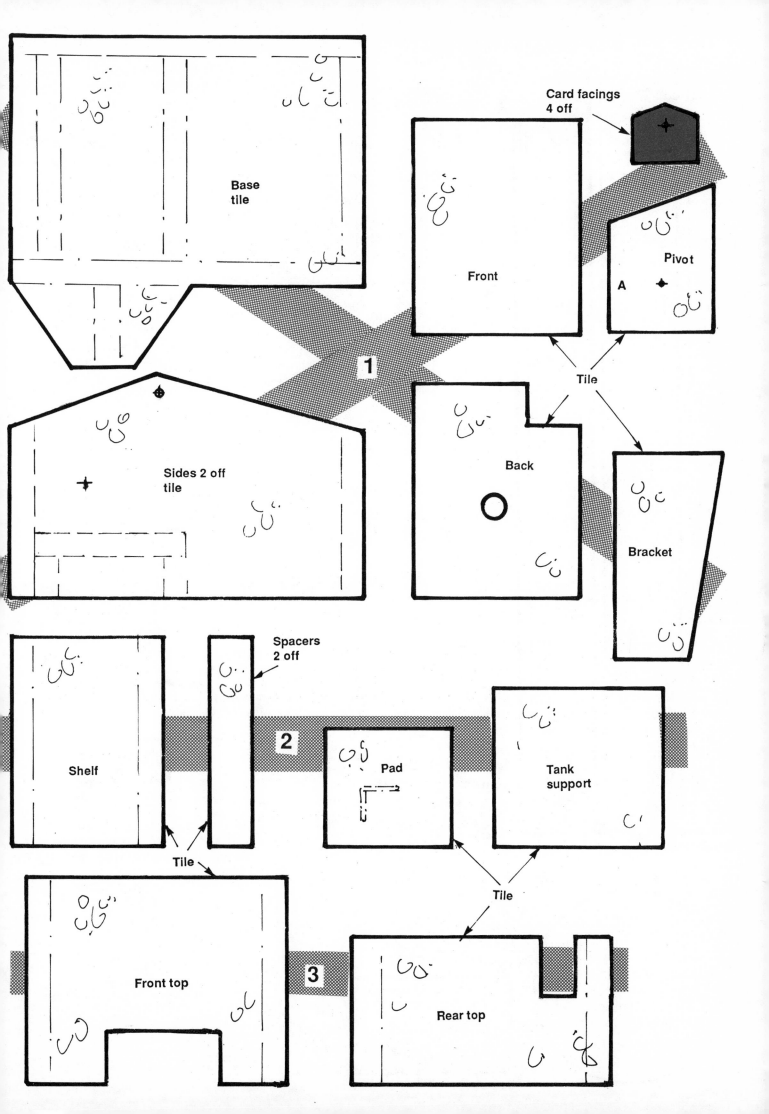

Base
tile

Card facings
4 off

Front

Pivot

A

Tile

Sides 2 off
tile

Back

Bracket

1

Spacers
2 off

Shelf

Tile

2

Pad

Tank
support

Tile

Front top

Rear top

Tile

3

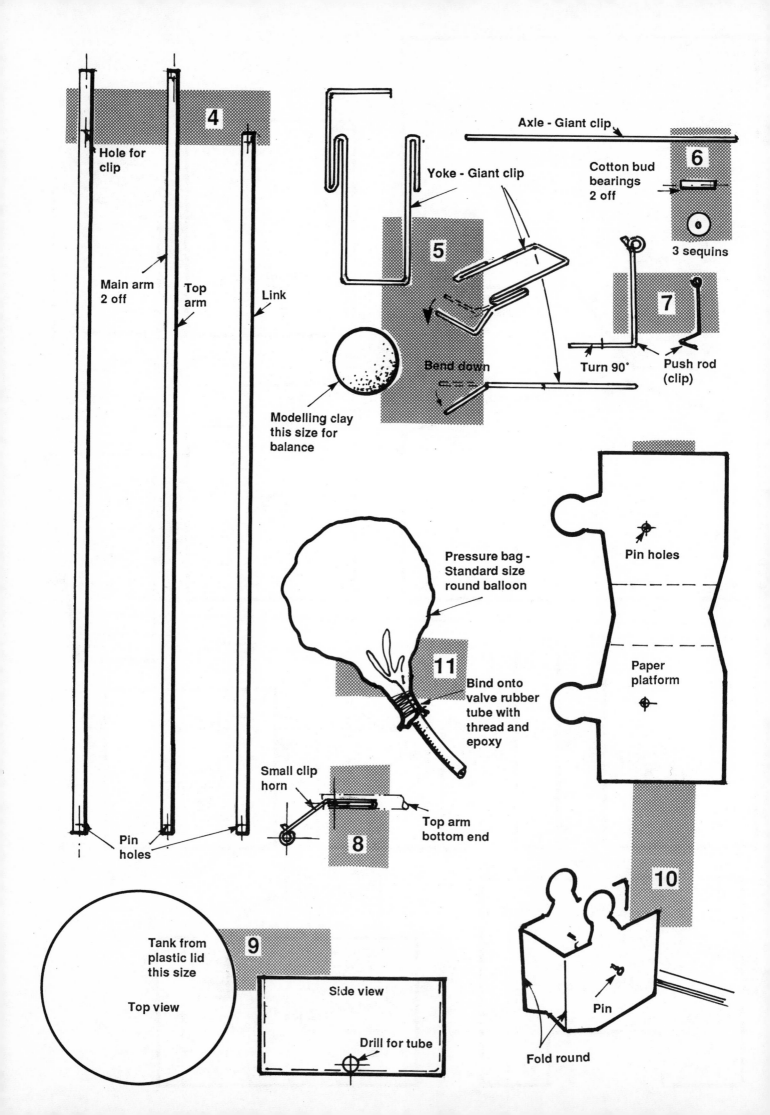

4

Hole for clip

Main arm 2 off

Top arm

Link

Pin holes

5

Yoke - Giant clip

Modelling clay this size for balance

Bend down

Axle - Giant clip

6

Cotton bud bearings 2 off

3 sequins

7

Turn 90°

Push rod (clip)

Pressure bag - Standard size round balloon

11

Bind onto valve rubber tube with thread and epoxy

Pin holes

Paper platform

Small clip horn

Top arm bottom end

8

9

Tank from plastic lid this size

Top view

Side view

Drill for tube

10

Pin

Fold round

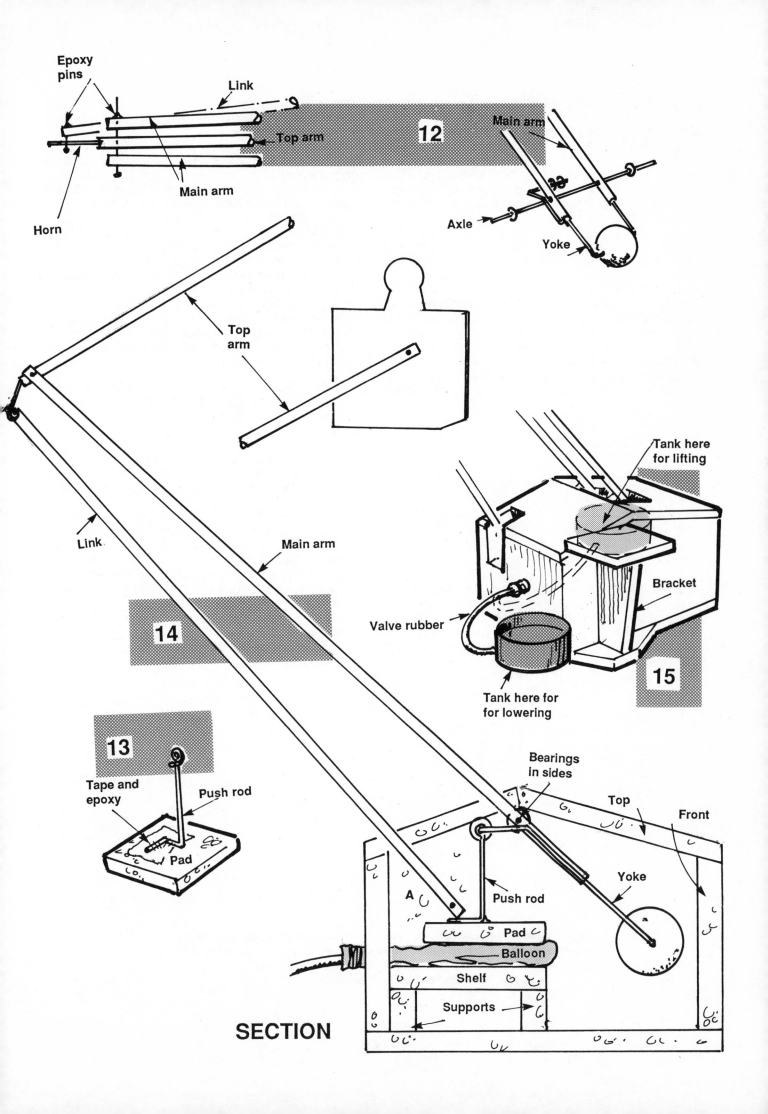

Epoxy
pins

Link

Top arm

12

Main arm

Main arm

Horn

Axle

Yoke

Top
arm

Top
arm

Tank here
for lifting

Link

Main arm

14

Bracket

Valve rubber

15

Tank here for
for lowering

Bearings
in sides

13

Top

Front

Tape and
epoxy

Push rod

A

Push rod

Yoke

Pad

Pad

Balloon

Shelf

Supports

SECTION

Step 6 Straighten a giant clip and use it as the main arm axle. Cut a cotton bud stem to make two bearings and epoxy these into the side pieces (Step 1). Sequins will form washers later.

Step 7 Form a pushrod from another clip. The looped end has to be formed over another clip to get it small enough. Make a double 90° bend at the bottom end.

Step 8 Bend a small clip to form a drive horn to fit into the bottom end of the top arm straw. Note the angle.

Step 9 Cut the hair spray, or similar-sized, lid down to make a shallow tank of the size shown. Pierce it at the corner to fit the valve rubber.

Step 10 Cut the platform, complete with silhouettes of passengers, from paper and fold it into shape. Pierce it to pivot loosely on a pin, but so that it hangs level.

Step 11 Blow up the balloon, leave it for a while, then untie it and let it down. It will be more flexible and slightly larger than before it was blown up.

Cut most of the neck away. Place what remains of the neck, still on the balloon, around one end of the valve rubber tube. Smear lightly with epoxy and bind with thread to make a tightly folded and watertight joint. Check that air can be blown gently in. When it's set, squeeze the balloon flat.

Step 12 Push pins through the straws to join and pivot the top arm to the main ones. Another pin goes through the top arm horn into the link arm. The axle goes through the main arms and has sequins on each end to centre it in the bearings. Epoxy the main straws to the axle.

Step 13 Epoxy the pushrod to the pad and add masking tape to steady it. Hook it onto the yoke and centre it between sequins.

Step 14 With the base box epoxied together, and the arms pivoted to it, pin the link arm to pivot freely exactly where shown. The bearings and this pin should go through card facings to reinforce the tile at these points.

The spacing of the link pivot from the axle, its angle and the angle and length of the top arm horn, all combine to make the top arm move at the right angle to the main arm, because its top end goes up faster than its lower end. This whole arm assembly must be kept very light, so do not paint it or make an elaborate job of the passenger platform. Thread the balloon into the space between the shelf and the pad so that it is flat and does not project each side. Do not fit the top covers yet.

Step 15 Plug the valve tube into the tank. Fill the tank with water and collapse the arm gently to a few inches from the top of the box. Air bubbles may come out.

Place the tank on the tank support. The water should flow into the balloon and expand it. This should force the pad and pushrod up, lifting the main arm. As it rises, the link arm will pull the top arm up too.

Add more clay to the yoke if the arm does not rise, or if it rises more slowly than it descends when the tank is replaced on the floor. Remove some clay if the arm goes up faster than it comes down, or if it stays up. At rest, with the arm down, the water level in the tank must be below the neck of the balloon as it projects out of the back of the box. The amount of water in the tank will also slightly alter the 'up' and 'down' speed.

When adjusted, pin the top covers in place, so that access is possible.

NOTE:
All the models described in this book have been built from the drawings and accompanying text, using only the recommended materials and assembly notes. The success of their operation will depend on the readers adherence to the instructions and his/her individual ability.

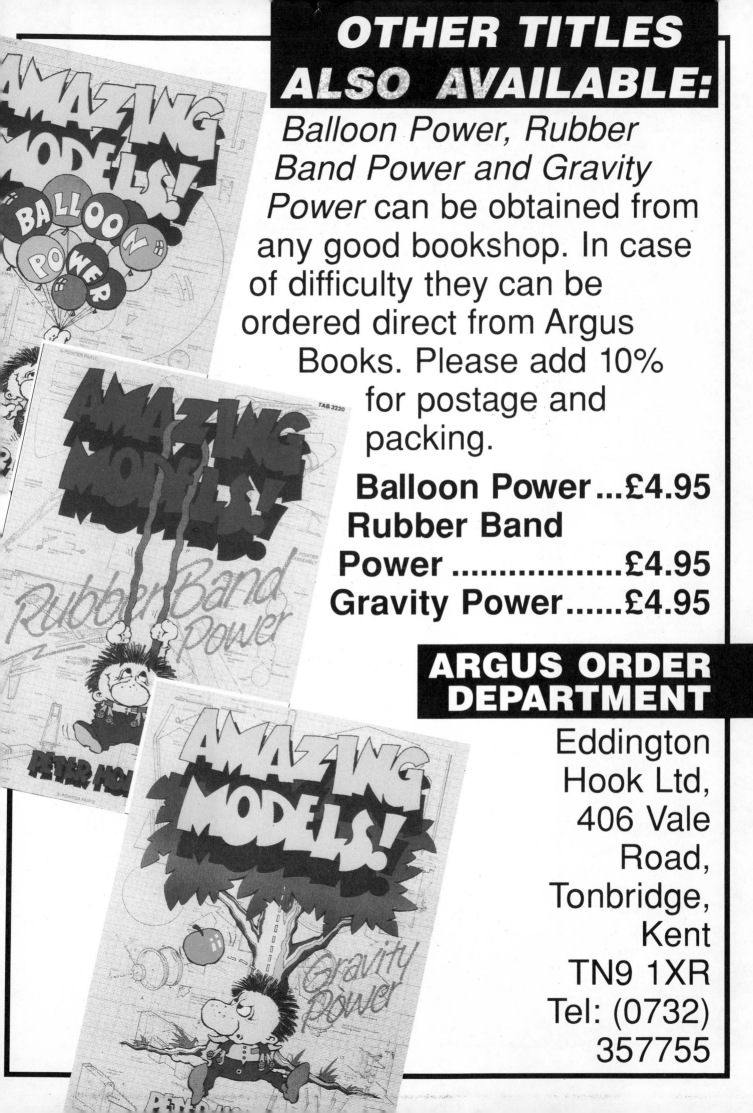